15th International Symposium on Electron Molecule Collisions and Swarms 2007

Journal of Physics: Conference Series Volume 115

Berkshire, United Kingdom
2 – 4 August 2007

ISBN: 978-1-5108-0426-5
ISSN: 1742-6588

Printed from e-media with permission by:

Curran Associates, Inc.
57 Morehouse Lane
Red Hook, NY 12571

Printed by Curran Associates, Inc. (2015)

For permission requests, please contact the Institute of Physics
at the address below.

Institute of Physics
Dirac House, Temple Back
Bristol BS1 6BE UK

Phone: 44 1 17 929 7481
Fax: 44 1 17 920 0979

techtracking@iop.org

Additional copies of this publication are available from:

Curran Associates, Inc.
57 Morehouse Lane
Red Hook, NY 12571 USA
Phone: 845-758-0400
Fax: 845-758-2634
Email: curran@proceedings.com
Web: www.proceedings.com

TABLE OF CONTENTS

Journal of Physics: Conference Series **115** (2008) 012001 doi:10.1088/1742-6596/115/1/012001

Space-time resolved kinetics of low-pressure breakdown

D Marić, G Malović, M Radmilović-Rađenović and Z Lj Petrović

Institute of Physics Belgrade, Pregrevica 118, 11080 Belgrade, Serbia

E-mail: draganam@phy.bg.ac.yu

Abstract. A review of diagnostics of low-current low-pressure discharges is given with an aim to illustrate how such discharges are used to determine swarm parameters and also how such data may be applied to model and understand the discharges. We have revised how comprehensive modelling of breakdown has led to agreement between binary collision data and the data that may be inferred from the breakdown (Paschen) curves by including processes such as space charge (current) effect on the local field in front of the cathode, photoemission, heavy particle gas phase ionization and backdiffusion. It is also discussed how modelling of Volt-Ampere characteristics in addition to Paschen curves is necessary to establish models of secondary electron emission and how these models may be applied in high current discharges. Finally we show how space time resolved anatomy of the breakdown can lead to understanding of the physics of the initial stages of gas breakdown and formation of Townsend regime, glow and abnormal glow discharges.

1. Introduction

Conditions for definition of electron swarms are that electrons should not perturb the external field, should not be affected by Coulomb forces due to either other electrons or ions or should not have collisions that affect their properties with molecules excited by other swarm particles. Such conditions are found in non-selfsustained and also in Townsend dark discharges. In addition, in other regimes of discharges, conditions may be found where swarm parameters may be applied directly and locally. Townsend regime discharges serve both as a source of swarm data and also may be directly described by swarm parameters through application of Townsend's theory of breakdown. With significant reduction in number of swarm experiments transport data are more often obtained from gas discharges. In addition to Townsend's discharges, early stages of gas breakdown are often associated with swarm physics and described by direct application of swarm parameters. In this paper we apply novel diagnostics of Intensified CCD cameras (ICCD) to study the development of low pressure discharges through breakdown and different regimes.

Phelps and coworkers [1-5] have recently started a systematic revision of Townsend's theory of low pressure breakdown and low current dark discharges (Townsend discharges), as it could not explain some phenomena that were observed in experiments [1-5], such as negative differential resistance, oscillations and discharge constriction. Most of the elements in their recent work have been presented previously by other authors; however, this has been a comprehensive study including all important processes. This study is based on quantitative comparisons with experimental data. By including space charge effects [1-3] and feedback mechanisms other than ion induced production of electrons at the cathode [5], it became possible to develop corrections to the standard Townsend's breakdown theory in order to explain negative differential resistance and its relation to oscillations in low current diffuse regime [1-4]. Consequently, the transition to the constricted regime became subject of further studies [4,6,7].

Systematic, well defined measurements in the dark Townsend regime [1-4,7] led to reanalysis of the role of different elementary processes in electron/ion production in gas breakdown and in the maintenance of self-sustained discharges [5]. In particular, it has been established that the secondary electron yield at the surface of the cathode depends on the mean energy of ions hitting the surface. The ion energy is affected by the external voltage, by the local reduced electric field E/N (where E is the electric field and N is the gas number density) and also by the current which leads to a perturbation of the field mostly in front of the cathode where ion concentration is the highest. It has also been shown that this weak dependence of the secondary yield on the space charge perturbation to the field, associated with a strong dependence of the ionization coefficient on E/N near the operating point, causes negative differential resistance that has been observed in low-current Townsend discharges [1-4,7]. Extension of Townsend's theory developed on the basis of the local-field approximation by Phelps and coworkers [1-3] was shown to be quite successful and it gave a good agreement with the experimental results for field distributions [7].

Extending these studies to somewhat higher currents and transition to constricted (normal glow) and diffuse (abnormal) glow discharges one may utilize the Volt-Ampere characteristics or the spatial profiles of emission to test the models of non-local electron transport and modelling of secondary electron yield that is to be applied for higher current discharges [8,9].

Our most recent studies in this field include time and space resolved development of different regimes of low-pressure discharges. Time resolved development of gas discharge structure is related to the kinetics of basic processes that participate to low-pressure breakdown and discharge maintenance. Similar studies have been recently reported – studies of spatiotemporal development of cathode-fall dominated DC discharges in argon [10,11] and of ignition of high current glow discharge in complex (parabolic) geometry [12]. Our aim was to extend the investigation to different modes of low-pressure discharges and to extend the knowledge of kinetics of formation and maintenance of these discharges in a simple geometry. We have followed the transition from steady-state Townsend to glow discharge and indications of the development of the space charge effects. This approach enables us observe very directly processes leading to development of the constriction, cathode fall and other features of glow discharges. The discharge in glow regime does not satisfy the conditions for swarm physics but many models are based on implementation of basic transport coefficients and our measurements may reveal why and how such approximations may fail or work.

2. Experimental setup

In our experiment, the discharge is established in a simple plane-parallel geometry. The cathode is made of copper, while the anode is made of quartz with transparent yet conductive thin film of platinum deposited on its surface. Thus it is possible to record both radial and axial profiles of emission. The diameter of the electrodes is 5.4 cm while the electrode separation can be set at three different values – 1.1 cm, 2.1 cm and 3.1 cm.

We run a continuous very low current (1-2 μA) discharge by applying a DC voltage to resistors connected in series with the electrodes. Triggering part of the circuit produces short current pulses superimposed on the DC voltages and currents. Pulses of higher current usually last only long enough ($\Delta t < 2$ ms) to make a reliable recording of voltage and current transients. The voltage is measured by two probes, one at the cathode and the other in the anode circuit. The second probe is used only when a relatively high monitoring resistor is connected into the low-voltage anode circuit to determine the current. The emission profiles recorded by the ICCD camera correspond to the conditions of the pulse in the observed time interval. Keeping the pulses short enough enables us to avoid heating and significant conditioning of the cathode surface during the measurements.

The delay generator built into the ICCD camera (Andor, iStar DH720-18U-03) enables us to synchronize recording of the light emission with pulse development and voltage-current measurements.

The details of experimental setup are presented in our previous papers (e.g. [8,9]).

3. Recent developments in description of Townsend's discharges

In this section we will review and discuss some of the new results in low-pressure gas discharge studies. Basic and, at the same time, the most questionable assumptions of the standard Townsend's theory (at least the way it is presented in the standard textbooks) are that secondary electron emission is induced only by ions hitting the cathode surface, electric field is homogeneous and that hydrodynamic conditions are satisfied in the discharge (local field approximation applies everywhere). On top of that several processes occurring at surfaces may affect the discharge very strongly.

3.1. Secondary electron yields in homogeneous field

For a long time, there was a great deal of confusion in the literature about the use of secondary yield data in breakdown modelling and, generally about the meaning of the data entering the breakdown condition. The secondary electron yield data obtained from swarm experiments have always failed to match the direct measurements in the binary beam-surface experiments. Phelps and Petrović [5] have shown that in the case of argon, the well known, almost constant, secondary yield of around 8 % for argon ions that has been obtained by ion beams on surfaces cannot be directly applied to model even the basic low pressure breakdown. A large number of processes participate in secondary electron production at the cathode (see [5] and references therein). As a matter of fact, ions dominate only in a very narrow range of moderately high *E/N*. Secondary electrons are produced mainly by photons at low *E/N* and heavy-particle ionization of gas atoms at high *E/N*. Also, metastables have observable contribution at all *E/N*. Including all these processes with appropriate experimental collision data came close to actually reconciling the results for secondary yields obtained in beam experiments and by analysis of the Paschen curve or other gas discharge techniques [5].

Reliable secondary electron yield data obtained from gas breakdown (i.e. by application of Paschen's law) are required to complete the understanding for other gases. We have analysed Paschen curves and spatial emission profiles from our measurements for a number of gases [13,14]. One has to be aware of two main problems when applying breakdown data. Spatial emission profiles may be used to take into account the real ionization rate [13]. However, hydrodynamic conditions are not completely satisfied in discharge, as it is assumed in Townsend's theory. There is a distinct non-hydrodynamic region close to the cathode (figure 1). Not taking into account this region can lead to significant errors. In figure 2, we show that discrepancy in secondary electron yields in argon can be up to an order of magnitude if d_0 is not taken into account. At high *E/N*, the multiplication becomes smaller, close to 1 because of the low pressure, so the two curves converge as inclusion of the non-equilibrium cannot make a significant change in the multiplication. At high pressures the curves converge because d_0 is small and thus its inclusion makes little change in the overall multiplication. A very good agreement is achieved between the measured values of d_0 and those determined by the Monte Carlo code with realistic cross sections [14], at least for atomic gases. The semiempirical formula has also been proposed in [5]. Thus, for experiments that do not allow measurement of the spatial emission profiles, calculated data may be used.

The second problem is that the ionization rate taken from the literature may give a quite different multiplication (slope) as compared to the actual experiment and even small errors in ionization coefficient result in large discrepancies of the secondary electron yield [13,14]. This is in general a part of treatment of non-hydrodynamic swarms and one needs to be aware that most of the charged particle kinetics close to the electrodes will proceed in a non-local fashion, which affects not only the discharge itself but also many of the boundary processes such as back diffusion [15] and reflection [16,17].

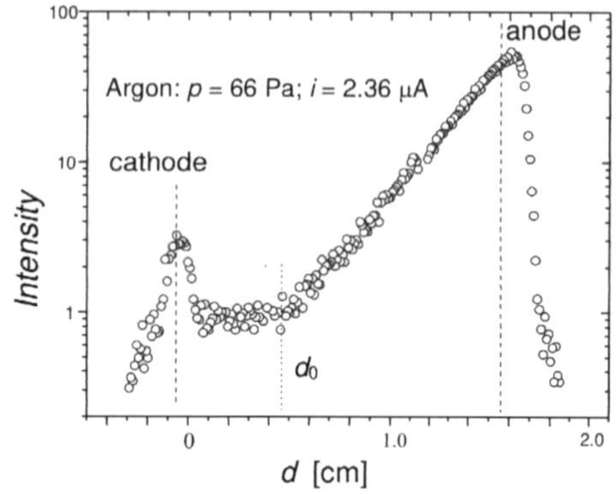

Figure 1. Axial profile of emission in low-current Townsend discharge. The width of the non-equilibrium region (delay length) is marked as d_0.

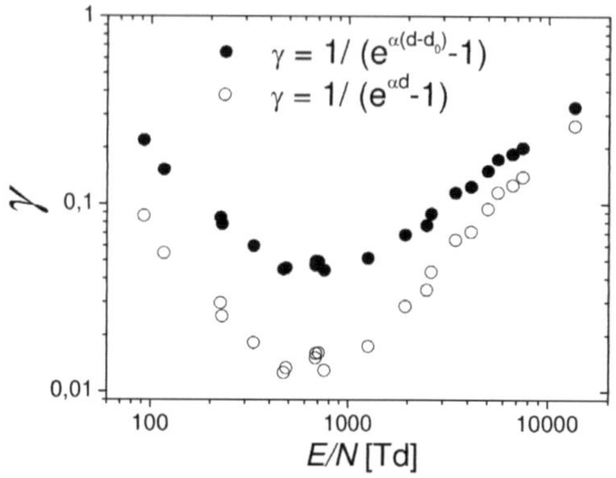

Figure 2. The secondary electron yield for breakdown in argon obtained from Paschen curves when relaxation length is: included (solid symbols) and not included (open symbols) in the analysis.

3.2. Space charge effects: negative differential resistance and oscillations

A study of the low current diffuse discharges revealed that the origin of the negative differential resistance, and consequently oscillations, is the perturbation of the electric field and its effect on the effective ionization coefficient and secondary electron yield [1-3,7]. The non-uniformity of the field is due to the build-up of the space charge increasing the field close to the cathode and thus changing the value of ion induced electron yield. Moreover, it has been shown that even at very low currents these effects are observable. As the matter a fact, phenomenology of dark Townsend discharge and region that is usually labelled as "subnormal" is the same the difference being in the degree of space charge effects. In discussing the values of frequency and the damping of the oscillations we should distinguish between the different types of ions in the discharge and the main mechanisms of secondary electron yield on the cathode.

It has been shown that secondary electron yields obtained for homogeneous field conditions can not be used in modelling of higher current discharges where space charge effects become notable [18]. Nikolić *et al.* [18] have shown that secondary electron yield dependence on electric field (discharge current) has to be taken into account in order to get physical solution of the model (negative slope of voltage-current characteristics in low-current regime).

Theoretical descriptions of the transition from the low current diffuse to the constricted (normal) glow include those trying to predict the shape of the channel in imposed or self-consistently calculated field distributions by adding the radial fields and diffusion losses. However, the onset of transition is determined by processes at low currents. Clearly, Townsend's theory in its basic form is not sufficient to achieve a totally adequate description, but some corrected forms of this theory may prove to be sufficient. Theories such as that of Phelps and coworkers [1-4] may be extended to predict the onset of transition to constrictions and its principal causes even without attempting to predict the exact properties of the constricted discharge itself.

3.3. Glow discharges

Significant advances in discharge modelling have been made in the past years, especially with the development of hybrid codes (e. g. [8,9,11,19]). Hybrid models are able to deal with the spatial non-locality of the electron transport in the cathode sheath and have provided important information about the phenomena taking place in a wide variety of discharge.

Hybrid codes, usually assuming a constant value for γ, have been successfully applied in studies of glow discharges [19]. In order to further improve the accuracy of these models possibility of use of secondary electron yields obtained for low-current discharge conditions (i.e. homogeneous field) [5] would yield satisfactory results in the region of the normal (constricted) glow and abnormal (high current diffuse) glow was investigated [20]. This test, however, resulted in a non-physical behaviour of the electrical characteristics (negative differential resistance in the abnormal glow mode), mainly caused by different combinations and properties of feedback mechanisms resulting from significantly different field and space charge distributions.

As an alternative approach, secondary electron yield has been used as a fitting parameter in some of the recent studies, it has been adjusted to fit different discharge conditions and to match the calculated electrical characteristics with the experimental results [8,9]. The validity of the model has been tested on the basis of our experimental results. The values obtained in this study have been found to be significantly lower compared to the case of the homogeneous field [5] and to change remarkably with changing discharge conditions. This implies that it is generally difficult to prescribe a certain value of γ which describes the discharge correctly for a wide range of operating conditions. Using energy-dependent secondary yields, for argon ions and fast atoms and following the procedure proposed by Phelps and Petrović [5], a good fit of the normal and abnormal glow discharges has been achieved [8,9].

Application of the prescription of Phelps and Petrović [5] in higher current discharges and in RF discharges (where phenomenology of gas breakdown may be quite different) appears to be difficult as it is derived on the basis of the breakdown and low current limit where linearity between electron flux and other fluxes holds. This may not be the case in higher current discharges with complex field profiles. Nevertheless the fact that it worked well for DC glow discharges means that modelling of secondary electron yields need not be more complex than that of Phelps and Petrović [5].

4. Voltage-current characteristics and spatial profiles of emission – steady-state measurements

Paschen curve in the range of conditions relevant for our study is shown in figure 3. We have performed measurements in argon at pressure (*p*) × electrode gap (*d*) products 250 Pa cm, 150 Pa cm and 45 Pa cm and *d* = 1.1 cm, 2.1 cm and 3.1 cm. *pd* products that are close to Paschen curve minimum (*pd* = 150 Pa cm) were used as a test case. Higher *pd* values were interesting because of the constricted modes of a discharge, and lower *pd* values to study contribution of heavy particles to discharge operation.

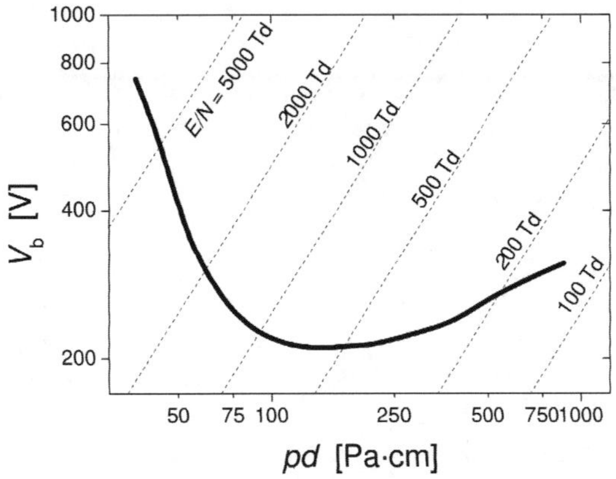

Figure 3. Breakdown voltages as a function of gas pressure times the gap length (*pd*) (Paschen curve) for argon.

Our study covered formation and maintenance of different modes of discharges – low current diffuse (Townsend) discharge, normal glow and abnormal glow discharge. Figure 4 illustrates the relation between voltage-current characteristics and the anatomy of the discharge for typical modes of low-pressure discharges. In low-current Townsend regime, axial profiles of emission exhibit exponential increase of emission from the cathode towards the anode. At the same time, radial profiles show that discharge is diffuse. Increase of current will lead to gradual change in exponential axial profile due to space charge formation. Eventually, at the transition from low-current regime to the normal glow, space charge effects will result in formation of cathode fall, which is revealed through the development of the peak of emission. As the current increases, the peak moves towards the cathode. Radially, the discharge structure can be significantly constricted. Constriction broadens up with the increase of current, until it occupies the whole electrode area at the beginning of abnormal glow regime of discharge.

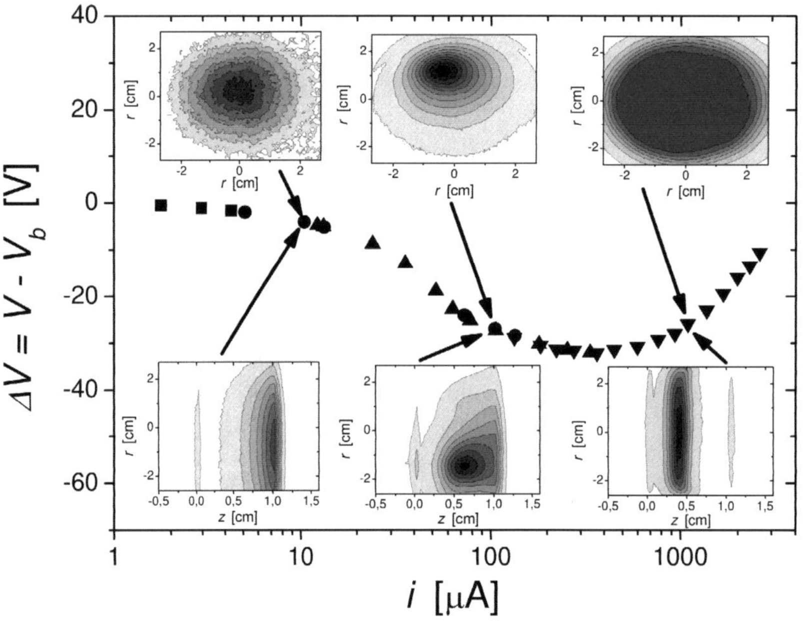

Figure 4. Volt-Ampere characteristics of low pressure discharge with spatial structure of the most characteristic discharge regimes indicated. Upper insets show radial contour plots of Townsend discharge, constricted glow discharge and abnormal glow discharge. Lower insets show corresponding axial structure of the discharge (cathode is positioned at $z = 0$, while anode is situated at $z = 1.1$ cm).

5. Time-resolved measurements of low-pressure discharges development

The basis of these results are temporally resolved images of discharges made by a fast ICCD camera, supported by voltage-current measurements. Discharge is established in a plane-parallel system of electrodes and it operates in a pulsed DC regime with variable peak currents. We have performed measurements that covered formation and maintenance of different modes of discharge – low current diffuse (Townsend) discharge, constricted normal glow and abnormal glow discharge. Special care was taken in recording of the spatiotemporal development of discharge oscillations and constrictions. The development of glow and abnormal glow discharges was recorded in side-on and end-on view, revealing radial dependence of the discharge and showing development of constrictions. We selected to present here one of the most interesting cases – formation of constricted (normal) glow discharge. Temporal development of other modes of low-pressure discharges will be presented elsewhere [21].

Development of constricted form of discharge was recorded at pd = 250 Pa cm. Under these conditions radial constriction is well developed, while negative differential resistance in low-current limit is low enough to achieve stable and reproducible results. A voltage-current characteristic for 250 Pa cm is shown in figure 5. Figure 6 shows current and voltage waveforms that correspond to development of constricted regime of discharge. Steady-state values of discharge voltage and current are indicated by an arrow in figure 6. 2D images of the discharge are presented in figure 7.

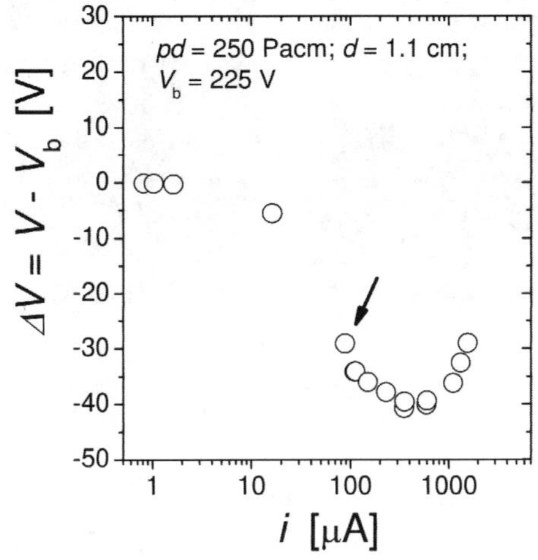

Figure 5. Voltage-current characteristics for pd = 250 Pacm and d = 1.1 cm. Temporal development of discharge is described for selected steady-state conditions, indicated by the arrow [21].

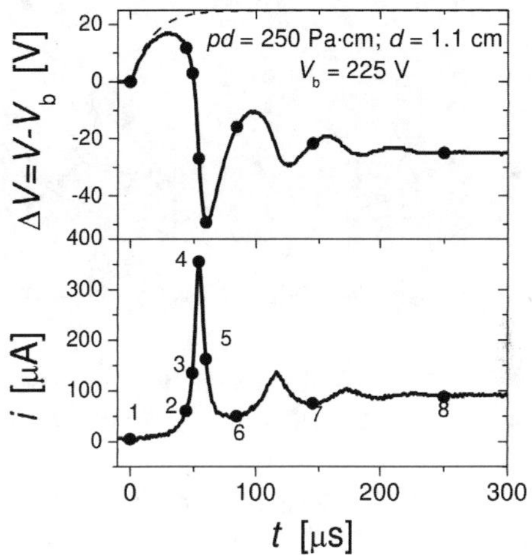

Figure 6. Current and voltage waveforms during formation of discharge constriction. Final steady-state values of voltage and current correspond to conditions indicated by an arrow in figure 5 [21].

Starting from Townsend's diffuse regime and typical exponential increase of intensity towards the anode, the discharge gradually exhibits space charge effects as the current increases. Gradual formation of the peak of emission can be observed, which is consistent with the formation of the cathode fall [7-11]. The peak of emission rapidly moves towards the cathode. At this point, the discharge is centred and curved towards the cathode (label 3 in figure 7), which is typical for rapid cathode fall development (rapid decrease of voltage, followed by the current growth). Detailed analysis of this kind of discharge development will be analyzed further on in text.

As the discharge current reaches maximum, the radial profile of the discharge broadens and the peak intensity moves further towards the cathode (label 4). Following the decrease of the current, the intensity of emission decreases and the peak of emission moves away from the cathode (label 5), until Townsend-like profile develops (label 6). As the discharge approaches the steady state, formation of constriction can be observed. During this phase of discharge development, the peak of emission gradually moves axially towards the cathode and radially towards the dielectric wall of the discharge chamber.

As it has been observed, throughout the fast cathode fall formation a radial deformation in emission spatial structure occurs (figure 7 1-8). Our initial assumption was that this kind of behaviour is due to the space charge effects. The basic idea was that the discharge can be considered as a group of independent parallel channels. In order to check this assumption, we followed development of spatial structure of the discharge in three different channels of the discharge. We have selected channel of the discharge that corresponds to the maximum intensity at the maximal current (axis of the discharge chamber) and channels that correspond to 90% and 70% of the emission intensity at the axis. Emission transients in different channels have been analyzed by establishing relation between position of the peak of emission d_{max} and peak intensity (figure 8).

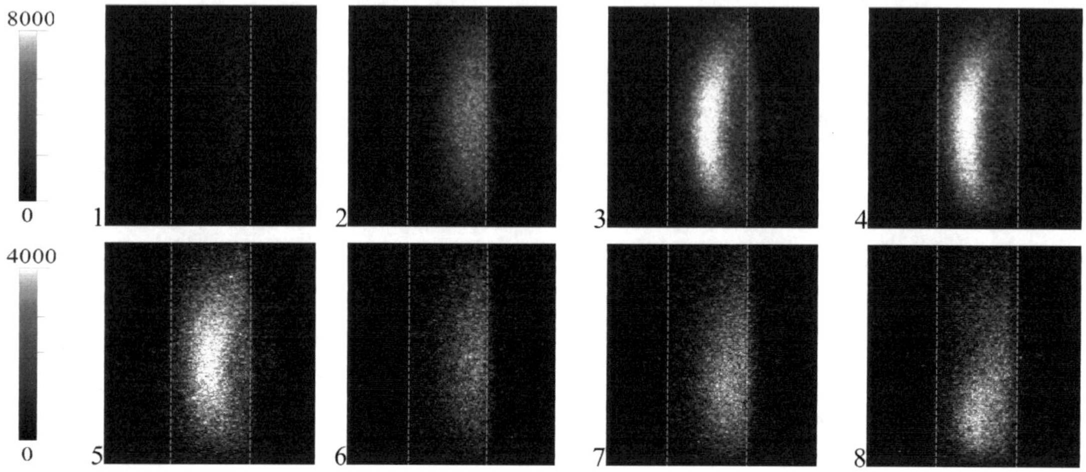

Figure 7. 2D scans of temporal development of discharge constriction. Labels 1-8 correspond to labels at figure 6. White dashed lines indicate positions of the cathode (left) and the anode (right) [21].

Figure 8 shows d_{max}(*max. intensity*) for selected channels of the discharge throughout the time interval indicated by the thick line in the current waveform (upper right inset). It can be observed that at specific times d_{max} are different for different discharge channels. However, all the channels generally follow the same trend. Development of channels closer to the wall is delayed in comparison to the central channel. At the same time d_{max}(*max. intensity*) dependency exhibits hysteresis. This indicates non-equilibrium processes that occur during fast cathode fall development.

Presented results confirm our initial assumption that during the cathode fall development the discharge can be observed as a group of parallel channels that operate independently (or in weak dependence on diffusion coefficient).

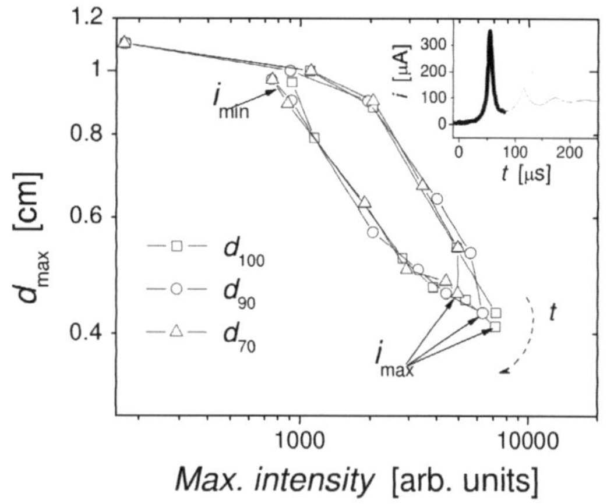

Figure 8. Position of peak of emission in dependence on peak intensity during cathode fall formation [21].

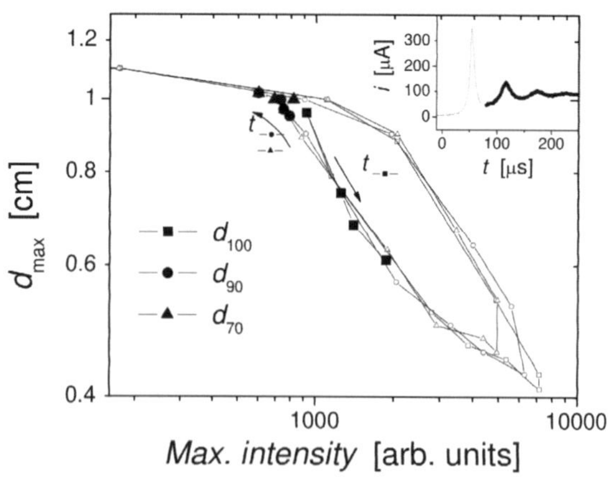

Figure 9. Position of peak of emission in dependence on peak intensity during formation of constriction [21].

In order to investigate further creation of constriction we will continue with the analysis of the development of the spatial structure of the discharge along the same current-voltage waveform. Figure 9 shows d_{max} for three selected channels of the discharge, for time interval indicated in upper right inset. We kept data obtained for cathode fall formation for comparison. During establishment of stationary state, in the channel that corresponds to highest emission intensity (solid squares) the discharge gradually develops from Townsend-like regime to normal glow regime of discharge operation. Peak of emission moves closer to the cathode, while intensity of emission increases. On the other hand, the remaining discharge channels retain Townsend-like behaviour with an exponential decrease of emission intensity.

Throughout formation of the constricted regime of the discharge, different discharge cannels become dependent. The current growth in one of the channels leads to turning off of the remaining channels due to the decreased operating voltage. Under the given operating conditions, electric field is too low for the discharge to operate in Townsend regime. This section of the discharge clearly operates in the non-selfsustained mode thanks to diffusion of charged particles from the constricted channel. Having this in mind, broadening of constrictions with an increase of the discharge current in the normal glow, can be explained. As the discharge current and charge particles density increase in the conducting channel, growing density of charged particles that diffuse to lateral channels will lead to an increase of the current in those channels and as a result to redistribution of the electric field and formation of the cathode fall. Eventually, the discharge will occupy the whole electrode diameter.

6. Summary

In this review we have covered some of the recent advances in description of the Townsend regime discharges and gas breakdown that is subject to Townsend's phenomenology and theory. These advances include a new comprehensive modelling of secondary electron emission due to ion, fast neutral, metastable, and photo emission as well as gas phase processes, inclusion of surface processes such as back diffusion and reflection, non-hydrodynamic transport of electrons (and other particles) close to surfaces, and many more processes.

In particular it was found that fitting of not only of the Paschen breakdown curve, but also of the Volt-Ampere characteristics is necessary to represent properly the kinetics of secondary electron emission.

At the same time measurements of properties of low current discharges which include both volt-ampere characteristics and spatial profiles of emission proved to be a fertile basis for modelling and our systematic measurements in different regimes of operation were analyzed by using a hybrid code and appropriate models of collisions, surface processes and in particular by using some aspects of secondary electron production as a fitting parameter.

In this paper we connect those studies with our recent measurements where space and time resolved recordings were made revealing some new phenomena such as curved front of emission and multi regime operation of the same discharge at the same time.

Analysis of normal glow spatial structure development in time enabled us to follow kinetics of cathode fall formation. A constricted form of discharge is established, as the discharge approaches the steady state. In particular it is important to note that coexistence of several modes of operation from Townsend to glow was observed in the same discharge at the same moment, depending on the local current density.

Our goal was partly to provide the experimental data that may be modelled by 2D models, which will provide further information on basic kinetic processes in the discharge. Nevertheless, based merely on inspection of the images one can reveal some new phenomena and settle some issues in the formation of gas discharges upon the breakdown of low pressure gases.

Acknowledgements
The work was supported by the project MNTRS 141025 and by COE IPB-CNP-026328. Authors are grateful to Dr. A.V. Phelps for many suggestions and to Drs. Z. Donko, K. Kutasi, P. Hartmann,

S. Živanov, I. Stefanović and N. Škoro for collaboration on the measurements and theory presented in this work.

References

[1] Petrović Z Lj and Phelps A V 1993 *Phys. Rev.* E **47** 2806
[2] Jelenković B M, Rózsa K and Phelps A V 1993 *Phys. Rev.* E **47** 2816
[3] Phelps A V, Petrović Z Lj and Jelenković B M 1993 *Phys. Rev.* E **47** 2825
[4] Petrović Z Lj and Phelps A V 1997 *Phys. Rev.* E **56** 5920
[5] Phelps A V and Petrović Z Lj 1999 *Plasma Sources Sci. Technol.* **8** R21
[6] Kolobov V I and Fiala A 1994 *Phys. Rev.* E **50** 3018
[7] Živanov S, Živković J, Stefanović I, Vrhovac S and Petrović Z Lj 2000 *Eur. Phys. J.* AP **11** 59
[8] Marić D, Hartmann P, Malović G, Donko Z and Petrović Z Lj 2003 *J. Phys. D: Appl. Phys.* **36** 2639
[9] Marić D, Kutasi K, Malović G, Donko Z and Petrović Z Lj 2002 *Eur. Phys. J.* D **21** 73
[10] Jelenković B and Phelps A V 1999 *J. Appl. Phys.* **85** 7089
[11] Donko Z 2000 *J. Appl. Phys.* **88** 2226
[12] Wagenaars E, Bowden M D and Kroesen G M W 2005 *IEEE Trans. Plasma Sci.* **33** 254
[13] Malović G, Strinić A, Živanov S, Marić D and Petrović Z Lj 2003 *Plasma Sources Sci. Technol.***12** S1
[14] Malović G, Marić D, Živanov S, Radmilović-Rađenović M and Petrović Z. Lj., unpublished
[15] Radmilović M and Petrović Z Lj 2000 *Eur. Phys. J. AP* **11** 35
[16] Vrhovac S B, Stojanović V D, Jelenković B M and Petrović Z Lj 2001 *J. Appl. Phys.* **90** 5871
[17] Radmilović-Rađenović M and Petrović Z Lj, in preparation
[18] Nikolić M, Đorđević A, Stefanović I, Vrhovac S and Petrović Z Lj 2003 *IEEE Trans. Plasma Sci.* PS **31** 717.
[19] Bogaerts A, Gijbels R, Goedheer W J 1995 *J. Appl. Phys.* **78** 2233
[20] Phelps A V, Pitchford L C, Pedoussat C, Donko Z 1999 *Plasma Sources Sci. Technol.* **8** B1
[21] Marić D and Petrović Z Lj, in preparation

Modified effective range analysis of low energy electron and positron scattering on CO_2

Zbigniew Idziaszek[1], Grzegorz P. Karwasz[2], Roberto S. Brusa[3]
[1]*Centrum Fizyki Teoretycznej, Polska Akademia Nauk, 02668 Warszawa, Poland*
[2]*Instytut Fizyki, Uniwersytet Mikołaja Kopernika, Toruń, Poland*
[3]*Dipartimento di Fisica, Università degli Studi di Trento, 38050 Trento, Italy*

Abstract. Analytical solutions for the modified effective range problem have been applied to positron and electron scattering on carbon dioxide in the low (below 10 eV) energy range. For positrons, the solution with three partial waves reproduces very well experimental results up to the positronium formation threshold; the *s*-wave contribution rises in the limit of zero energy and the *p*-wave contribution reaches a very broad maximum at about 0.5 eV. For electron scattering, the present solution shows a sharp rise of the *s*-wave contribution in the limit of zero energy, explained by earlier calculations as a virtual negative ion state. The *p*-wave shows a resonant structure at about 5 eV corresponding to an experimentally well known $^2\Pi_u$ shape resonance. An additional maximum in the *p*-wave contribution is observed at about 1-2 eV. The latter feature would explain resonant-like scattering observed recently in high-resolution vibrational excitation measurements [M. Allan, Phys. Rev. Lett. 87 (2001) 033201].

PACS: 34.85.+x, 39.90.+d

Correspondence author: zbyszek@cft.edu.pl

1. Introduction – experiments and theory

Knowledge of cross sections for electron scattering on CO_2 is of basic importance for understanding operation of high power infrared lasers; CO_2 is also the main factor in human-induced greenhouse effect. The low-energy electron scattering total cross section is characterized by a strong rise in the limit of zero energy, with the total cross sections reaching as much as [1] $60\text{x}10^{-20}\,\text{m}^2$ at 0.1 eV and presence of a shape $2\Pi_u$ resonance at about 3.8 eV, observed as a maximum in the total cross section [2,3], with a big part (about 1/3, see [4]) coming from the vibrational excitation.

While early calculations using a semiempirical polarization potential [5,6] agreed pretty well with the electron-scattering experiment in a large (0.1-10 eV) energy range, ab-initio theories covering this energy range have been developed only recently [7,8,9]. The low energy rise has been attributed [10,11] to a virtual Σ_g state. The recent measurements of the vibrational excitation showed the presence of resonant enhancements also in the 1-2.5 eV energy range [12].

Positron scattering on CO_2 was studied experimentally by few groups (see the discussion below). For the theory, Gianturco and Paioletti [13] applied the density-functional method for short-range interaction effects in order to obtain elastic and rotational excitation cross sections in the 0.5-7 eV energy range and Gianturco and Mukherjee [14] used the body-fixed vibrational close-coupling scheme to obtain the elastic and the vibrational excitation cross sections in the same energy range. The weak point of theoretical calculations is that different models were separately applied to positron and electron projectiles and for the very low and resonance energy regions in electron scattering.

The present work gives an analysis of positron and electron scattering on CO_2 in the framework of the modified effective-range theory (MERT) [15]. It is based on the analytical solution for scattering from a polarization potential with short-range effects introduced through parametrization of the phase shifts [16]. The analysis uses experimental total cross sections at energies below the threshold for electronic excitation (or positronium formation for positrons). In the this energy range (apart from resonances) the total cross section, measured with a few per cent error, approximates well the integral elastic cross section (usually measured with much bigger uncertainty and by normalized methods).

2. Choice of experimental data

For electron scattering the time-of-flight experiments in the very-low-energy range [1, 3] merge well with the electrostatic-beam experiment in the region of a few eV [2]. These three sets of data, weighted by their uncertainties form recommended values given in [4], are used for the present analysis.

Positron-scattering total cross sections were measured down to 0.5 eV by Hoffman et al. [17] with an apparatus using a long (109 cm) scattering cell, with rather small apertures (1.2 mm radius at the entrance and 2.4 mm at the exit) and a longitudinal guiding magnetic field (16 G at 2 eV collision energy) [18]. First measurements by Sueoka and collaborators [19] were performed down to 1 eV using 67.5 mm long scattering cell (we quote the geometrical length and not the "effective" 79.7 mm length) with 4 mm radius apertures and 9 G guiding field. The more recent data [20] (quoted also in [21]) were performed down to 0.3 eV with 1.8 G guiding field and 3 mm radius apertures; at 1 eV those data are by 10% lower than the value of Kwan et al. [22] and 20% higher than the early set [19].

The new apparatus constructed at Trento laboratory [23] uses a 10 mm long scattering cell with 0.75 mm apertures radii and 9 G guiding field. The positron scattering data for CO_2 extending down to 0.1 eV [24] rise quickly in the low-energy limit and form the highest set. As we discussed in detail for N_2 measurements [25], the use of large apertures in the scattering cell combined with high magnetic fields can lead to a strong underestimation of the total cross section in the low energy limit. At sufficiently low energies, when the cyclotronic radius of positrons in the magnetic field equals or is smaller than the scattering cell exit apertures, all positrons scattered into the forward cone (up to 90° scattering angle) are recaptured by the magnetic field and guided to the detector. Assuming an isotropic differential cross section and no inelastic processes present, the measured total cross section would then amount to 50% of the real value. Such a "half-value" collision energy equals to 0.8 eV in the experiment of [17], 1.1 eV in experiment of Sueoka and Mori [19], 0.026 eV in their repeated experiment [26], and as little as 0.0045 eV in the Trento experiments [25].

A second possible source of error in the absolute magnitude of total cross sections in the low energy limit can be the energy scale shift. In the Trento apparatus, detailed procedures both with a retarding field analyzer [27] and using the N_2 and Ar positronium thresholds [25] were applied, so a possible error in the energy scale is within ± 0.1 eV. Note also that the energy resolution of the Trento apparatus is about 150 meV, as proved by observation of sharp resonant structures in He at 1.4-2.1 eV [28]. For all these arguments we believe that the cross sections from Trento laboratory can be considered presently the most reliable set for positron scattering.

3. Analytical solution for modified effective range approximation

As already stated by O'Malley et al. [29] the zero energy limit in electron and positron scattering should be well approximated by the modified effective range theory. In this theory scattering is assumed to be due to polarization potential solely and the short-range effects are introduced by an effective range R_0.

Electron scattering on noble gases has been extensively evaluated in the framework of MERT by Buckman and Mitroy [30] who used five-parameter expansion in series of momentum k for the s and p-wave phase shifts. Electron scattering on non-polar molecules (H_2, N_2, CO_2) was studied, among others by Fabrikant [31]. In all these studies an upper limit for MERT analysis was confined to energies below 1 eV (0.2 eV for CO_2). In a recent work [16] we have applied analytical solutions for scattering on the polarization potential to extend the applicability of MERT to higher energies. We obtained MERT expansion by developing the parameter characterizing the short range interaction into series of k . We tested this scheme for positron scattering on Ar and N_2 up to 2 eV - the parameters of the short range potential were obtained by fitting recent experimental total cross sections [25].

We resume here the main notions. The radial Schrödinger equation for a particle moving in the polarization potential $V(r) = -\alpha e^2/2r^4$ is

$$\left[-\frac{\hbar^2}{2\mu}\left(\frac{\partial^2}{\partial r^2} - \frac{l(l+1)}{r^2}\right) - \frac{\alpha e^2}{2r^4} - E\right]\Psi_l(r) = 0, \tag{1}$$

where μ is the reduced mass, α is the static dipole polarizability, E is the relative energy of the particles, and $\Psi_l(r)$ denotes the radial wave function for the partial wave l. With appropriate change of variables, the Schrödinger equation (1) can be transformed into Mathieu's differential equation of the imaginary argument [32], and solved analytically in terms of the continued fractions (see e.g. [16]). At small distances where the polarization potential dominates over centrifugal potential and the constant energy term, the asymptotic behaviour of $\Psi_l(r)$ is described by

$$\Psi_l(r) \xrightarrow{r\to 0} r\sin\left(R^*/r + \phi_l\right) \tag{2}$$

where $R^* = e/\hbar\sqrt{\alpha\mu}$ describes a typical length related to the polarization interaction, and ϕ_l is some short-range phase, that depends on the short-range part of the potential. For l=0 the short range phase determines the s-wave scattering length: $a = - R^*\cot \phi_0$.

At large distances $\Psi_l(r)$ must take the form of the scattered wave

$$\Psi_l(r) \xrightarrow{r\to\infty} \sin\left(kr - l\,\pi/2 + \eta_l\right) \tag{3}$$

where $k = \sqrt{2\mu E}/\hbar$ and phase-shifts η_l can be expressed in terms of the short-range phases:

$$\tan\eta_l = \frac{m^2 - \tan^2\delta + (m^2-1)\tan\left(\phi_l + l\,\pi/2\right)\tan\delta}{(1-m^2)\tan\delta + \tan\left(\phi_l + l\,\pi/2\right)\left(1 - m^2\tan^2\delta\right)} \tag{4}$$

Here, $\delta = \pi/2(\nu - l - 1/2)$, and m and ν are some parameters characterizing Mathieu functions of the imaginary argument (see [16] for details). The effective range is introduced by expanding of $\tan(\phi_l + l\,\pi/2)$ in powers of k

$$\tan\left(\phi_l + l\,\pi/2\right) = A_l + \tfrac{1}{2}R^* R_l k^2 + \dots, \tag{5}$$

where $A_l = \tan(\phi_l + l\pi/2)\big|_{k=0}$ The lowest order correction in k is quadratic, and can be interpreted as effective range R_l for partial wave l [33].

4. Application for positron and electron scattering on CO_2.

Using for the (spherical) polarizability of CO_2 the value obtained from electron-scattering electronic excitation experiments α=16.92 a_0^3 [34] we obtain R^* = 4.113 a_0. For MERT fit for positrons we used all experimental data from Trento lab from 0.1 eV up to the positronium formation threshold (i.e. 6.8 eV). The experimental data can be approximated quite well assuming contributions just from two partial waves s and p. If MERT expansion is performed also for d wave, the agreement remains within the statistical error bar for the whole energy range considered, see figure 1. The s-wave partial wave cross sections rises in the limit of low energies and dominates over the p-wave contribution for energies below 0.2 eV; the s-wave contribution passes through zero at 2 eV. Parameters of the potential (zero-energy contributions A_l and the effective ranges R_l) for the three partial waves are given in table 1. Note the zero value for the p-wave zero-energy parameter A_1 and a negative value for the s-wave scattering length.

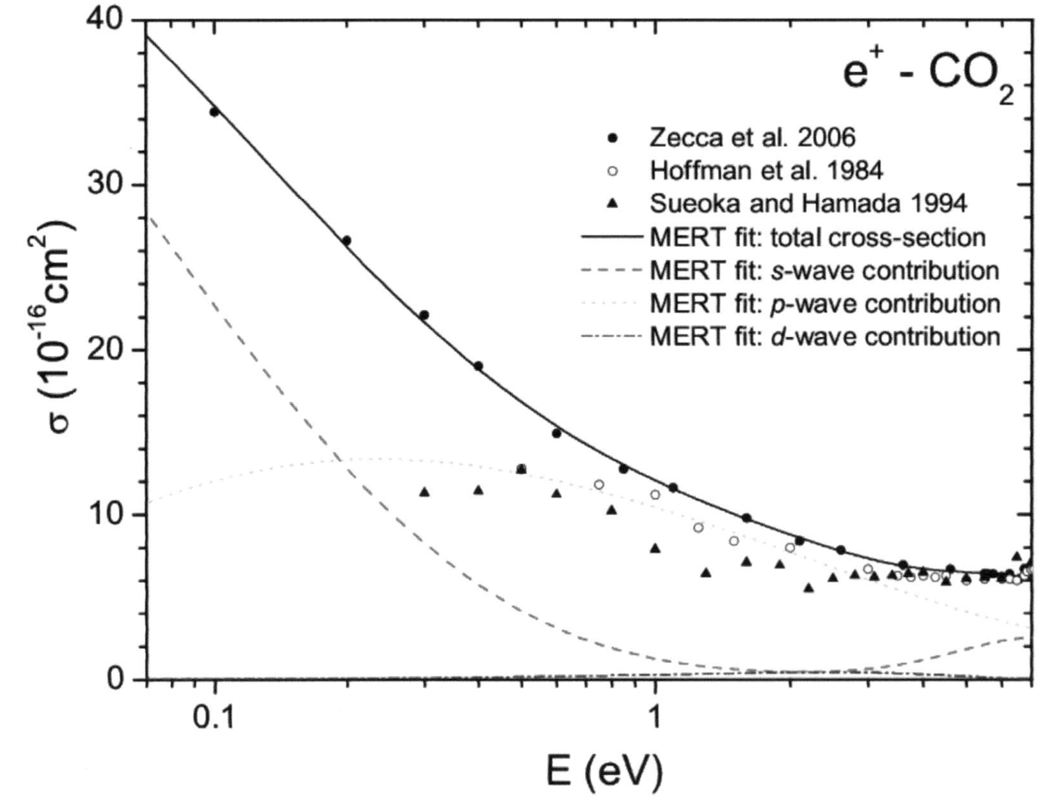

Fig. 1. MERT analysis for positron scattering on CO_2. Experimental data are: full circles – Zecca et al. [24], open circles – Hoffman et al. [17], triangles – Sueoka and Hamada [20]. Theoretical curves are: MERT fit for the total cross-section (solid line), and contributions from s wave (dash line), p wave (dot line) and d wave (dash-dot line).

	$a/R^*=-1/A_0$	A_1	A_2	R_0/R^*	R_1/R^*	R_2/R^*
$E^+ - CO_2$	- 1.120	- 0.0077	- 0.177	- 0.389	- 0.157	0.028
$e - CO_2$	- 1.616	- 6.250	- 0.406	- 0.266	2.780	0.131

Table 1. Positron and electron scattering parameters obtained by fitting MERT model to the experimental data.

For electron scattering we used the recommended data [4] from 0.1 to 2.0 eV: the fit with MERT expansion for two partial waves reproduced the set within experimental uncertainties, see fig.2. The obtained scattering length a =- 6.61a_0 agrees well with the value a=- 6.17a_0 predicted by Morisson [35] and the limits given by Fabrikant [31]: -7.2 $a_0 < a <$ -6.8a_0.

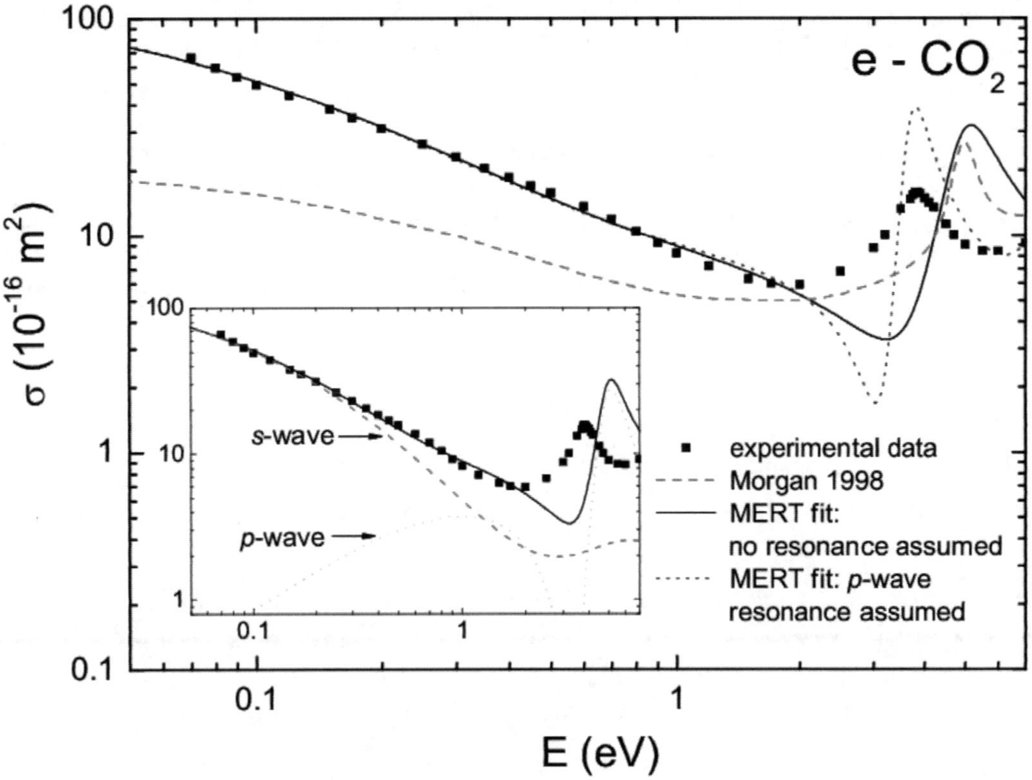

Fig. 2. MERT analysis for electron scattering on CO_2. Depicted are: recommended experimental data from review [4] (squares), multistate calculation of Morgan [11] (dash line), and theoretical fits with MERT expansion for two (solid line) and three (dot line) partial waves. For the latter calculation we assumed *p*-wave resonance at the experimentally observed peak. The inset shows in addition the *s*-wave and *p*-wave contributions to the MERT fit with two partial waves.

Differently from previous MERT approaches and in agreement with experiments, our MERT model predicts the existence of a shape resonance in *p*-wave channel at 5 eV. This is slightly higher than the peak in the experimental total cross section but also the calculation by Morgan [11] overestimates the energy of the shape resonance, see fig. 1. The shape resonance appears in the regime where the *d*-wave contribution can be significant, therefore we have also tried a fit with MERT for three partial waves fixing additionally the position of the resonance at the experimentally observed value. Parameters of such a fit are given for sake of comparison with positron parameters in table I. This demonstrates the compatibility of MERT expansion for small energies and the existence of the *p*-wave shape resonance.

Our MERT analysis shows also a peculiar behaviour of the *p*-wave shift just below the shape resonance. The *p*-wave cross section changes quickly just below the resonance, reaching another maximum at 1 eV. We hypothesize that this maximum is responsible for a resonance-like enhancement of the vibrational excitation observed between 1 and 2.5 eV by Allan [12].

6. Conclusions

We show that the recently developed scheme based on analytical solutions and MERT expansion gives very good agreement with experimental data both for positron and electrons. For positrons the agreement extends up to the threshold for positronium formation. For electrons, the fit performed in the low energy range (below 2 eV) predicts the existence of a shape resonance at 5 eV, and an additional enhancement in the *p*-wave channel at about 1 eV. The obtained position of the resonance is slightly higher than the experimentally observed peak at 3.8 eV, and is close to the theoretical value of [11]. The agreement with the experiment can be improved, as far as the position of the resonance is concerned, using three partial waves in MERT expansion. Summarizing, the present MERT calculations reproduces well the observed cross section for both type projectiles. In the case of positron scattering MERT shows a sharp rise of the total cross section in the limit of zero energy; in the case of electrons it shows a similar rise in the zero-energy limit plus a shape resonance at about 5 eV.

References

[1] J. Ferch, C. Masche, and W. Raith, J. Phys. B 14 (1981) L97.

[2] Cz. Szmytkowski, A. Zecca, G. Karwasz, S. Oss, K. Maciąg, B. Marinković, R.S. Brusa, and R. Grisenti J. Phys. B 20 (1987) 581.

[3] S. J. Buckman, M. T. Elfordt and D. S. Newman, J. Phys. B 20 (1987) 5178.

[4] G. P. Karwasz , A. Zecca and R.S. Brusa, Electron Scattering with Molecules, in: Photon and Electron Interaction, with Atoms, Molecules and Ions, Landolt-Börstein New Series vol I/17, Springer-Verlag, Berlin, Heidelberg, 2003, pp 6.1-6.51.

[5] M. A. Morrison, N. F. Lane, and L. A. Collins, Phys. Rev. A 15 (1977) 2186.

[6] D. Dill, J. Welch, J. L. Dehmer, and J. Siegel, Phys. Rev. Lett. 43 (1979) 1236.

[7] T. N. Recigno, D.A. Byrum, W. A. Isaacs, and C. W. McCurdy, Phys. Rev. A 60 (1999) 2186.

[8] C.-H. Lee , C. Winstead, and V. McKoy, J. Chem. Phys. (1999) 111 5056.

[9] T. N. Rescigno, W. A. Isaacs, A.E. Oral, H.-D. Meyer, and C. W. McCurdy, Phys. Rev. A 65 (2002) 032716.

[10] H. Estrada and W. Domcke, J. Phys. B 18 (1985) 4469.

[11] L. A. Morgan, Phys. Rev. Lett. 80 (1998) 1873.

[12] M. Allan, Phys. Rev Lett. 87 (2001) 033201.

[13] F. A. Gianturco and P. Paioletti, Phys. Rev. A 55 (1997) 3491.

[14] F. A. Gianturco and T. Mukherjee, J. Phys. B 30 (1997) 3567.

[15] L. Spruch, T. F. O'Malley, and L. Rosenberg, Phys. Rev. Lett. 5 (1960) 347.

[16] Z. Idziaszek and G. P. Karwasz, Phys. Rev. A 73 (2006) 064701

[17] K. R. Hoffman, M. S. Dababneh, Y.-F. Hsieh, W. E. Kauppila, V. Pol, J. H. Smart, and T. S. Stein, Phys. Rev. A 25 (1982) 1393.

[18] W. E. Kauppila, T. S. Stein, G. Jesion, M. S. Dababneh, and V. Pol, Rev. Sci. Instrum. 48 (1977) 822.

[19] O. Sueoka and S. Mori, J. Phys. Soc. Japan 53 (1984) 2491.

[20] O. Sueoka and A. Hamada, J. Phys. Soc. Japan 62 (1993) 2669.

[21] M. Kimura, O. Sueoka, A. Hamada, M. Takekawa, Y. Itikawa, H. Tanaka, and L. Boesten, J. Chem. Phys. 107 (1997) 6616.

[22] C. K. Kwan, W. E. Kauppila, S. Nazaran, D. Przybyla, N. Scahill, T. S. Stein, Nucl. Instr. Meth. B 143 (1998) 61.

[23] G. Karwasz, R.S. Brusa , M. Barozzi, and A. Zecca, Nucl. Instr. Meth. B 171 (2000) 178.

[24] A. Zecca, C. Perazzolli, N. Moser, D. Sanyal, M. Chakrabarti, and M. J. Brunger, Phys. Rev. A 74 (2006) 012707.

[25] G. P. Karwasz, D. Pliszka, and R. S. Brusa, Nucl. Instr. Meth. B 247 (2006) 68.

[26] O. Sueoka and S. Mori, J. Phys. B 27 (1994) 5083.

[27] G.P. Karwasz, R.S. Brusa , and D. Pliszka, Nucl. Instr. Meth. B 251 (2006) 520.

[28] G. P. Karwasz , D. Pliszka, A. Zecca, and R. S. Brusa, Nucl. Instr. Meth. B 240 (2005) 666.

[29] T. F. O'Malley, L. Rosenberg, and L. Spruch, Phys. Rev. 125 (1962) 1300.

[30] S. J. Buckman and J. Mitroy, J. Phys. B 22 (1989) 1365.

[31] I. I. Fabrikant, J. Phys. B 17 (1984) 4223.

[32] E. Vogt and G. H. Wannier, Phys. Rev. 95 (1954) 1190.

[33] T. F. O'Malley, L. Spruch, and L. Rosenberg , J. Math. Phys. 2 (1961) 491.

[34] T. N. Olney, N. M. Cann, G. Cooper, and C. E. Brion, Chem. Phys 223 (1997) 59.

[35] M. A. Morrison, Phys. Rev. A 25 (1982) 1445.

Investigating the sensitivity of nitric oxide infrared emissions to electron impact

L Campbell[1], M J Brunger[1] and M. Allan[2]

[1] ARC Centre for Antimatter-Matter Studies, SoCPES, Flinders University, GPO Box 2100, Adelaide, SA 5001, Australia

[2] Department of Chemistry, University of Fribourg, chemin du Musée 9, 1700 Fribourg, Switzerland

E-mail: laurence.campbell@flinders.edu.au

Abstract. Integral cross sections for low energy electron excitation of the 0→1, 0→2 and 0→3 vibrational modes in nitric oxide have quite recently become available [Trevisan *et al.* PRA **71**, 052714 (2005)]. In this study we adapt our recent work [Campbell and Brunger GRL, in press (2007)], to look at the effect of these new cross sections on the production of nitric oxide infrared radiation. Predictions from our model are compared with measurements from Espy *et al.* [Planet. Space Sci. **36**, 543 (1988)], with the inclusion of the new cross sections improving the agreement of the shape of the spectrum with the measurements.

PACS#: 34.80.Gs

1. Introduction

Nitric oxide is a minor constituent of the upper atmosphere. However, it is important because infrared radiation from nitric oxide is a significant cooling mechanism in this region. Thus computational models of the energy transfer processes in the atmosphere need to predict both the density of NO and the rate of infrared emission from it. We recently described an investigation of the contribution of electron impact to both the production of NO and its infrared emissions [1], particularly for pathways that have been neglected in other studies.

Electron impact produces NO indirectly by a number of mechanisms. The principal one is that photoelectrons (produced by sunlight) and auroral electrons dissociate N_2 into atoms that then take part in a chemiluminescent reaction with O_2. In this reaction most of the NO is produced in a vibrationally excited state and either radiatively decays, producing infrared emission, or is quenched in collisions. It is generally assumed that this chemiluminescent reaction, along with collisional excitation, accounts for all of the infrared emission from NO.

We have shown previously that other mechanisms due to electron impact may be important. The reactions of excited N_2 with O atoms are predicted to make a significant contribution to the NO density in both daytime and auroral conditions [2]. In our earlier work [1, 3] we employed low-energy electron impact cross sections for vibrational excitation of NO from [4, 5] that led for instance to predictions that electron impact makes a significant contribution to the $1 \rightarrow 0$ emission from NO. Here we reprise that earlier work by employing relatively recent vibrational (0→1, 0→2 and 0→3) integral cross sections [6] for electron impact excitation of NO. Those new cross sections include both theoretical complex-Kohn results, as well as experimental values derived by Trevisan *et al.* [6] from the excitation function measurements of Allan [7].

In this paper we compare our predictions to rocket measurements [8], in which all the electron-driven mechanisms need to be considered. The measurements are of infrared emissions produced in an aurora. As they depend on the NO density, this density is also calculated [1] for the conditions of the experiment. Based on this density the infrared emissions are predicted and then compared with the measurements. The sensitivity of the calculated results to various available integral cross sections [3, 6] is considered.

2. Sources of NO and emissions

NO in the upper atmosphere is produced principally by the reactions of ground and excited state nitrogen atoms, $N(^4S)$ and $N(^2D)$, with O_2 [9] :

$$N(^4S) + O_2 \rightarrow NO + O \qquad (R1)$$

$$\text{and } N(^2D) + O_2 \rightarrow NO + O\ . \qquad (R2)$$

In sunlight nitrogen atoms are produced by photodissociation of N_2 and reactions of ions produced by photoionisation. The latter creates photoelectrons which produce nitrogen atoms by dissociation and indirectly by further ion production. Nitrogen atoms and NO are similarly produced by electron impact by secondary electrons in aurora. These reactions, among others in a photochemical model, were used by Barth [9] to calculate densities of NO in the upper atmosphere.

Nighttime infrared emissions from vibrationally excited nitric oxide (NO^*) have been identified as due [8] to collisional excitation by O atoms (called nightglow),

$$NO(\nu' = 0) + O \rightarrow NO^*(\nu' = 1) + O \qquad (R3)$$

and the chemiluminescent reaction:

$$N(^2D) + O_2 \rightarrow NO^*(\nu' = 1 - 12) + O\ . \qquad (R2')$$

The similar reaction

$$N(^4S) + O_2 \rightarrow NO^*(\nu' = 1 - 9) + O\ . \qquad (R1')$$

has been omitted in some cases due to its much lower reaction rate [10] , but is included in the calculations below.

The measurements [4, 5] of integral cross sections for low-energy electron-impact vibrational excitation of NO

$$NO(\nu'' = 0) + e^- \rightarrow NO^*(\nu' = 1 - 3) + e^-$$

made it possible to consider this as a source of NO^*. Those measurements were combined into the "current" set of cross sections [3] that were used in a detailed simulation of NO infrared measurements [1].

More recently Allan [7] measured the cross sections for these transitions at 135°. These excitation functions were scaled by 4π for comparison with new theoretical values of Trevisan *et al.* [6]. While such a scaling is probably not physical, even at these low energies, due to the anisotropy of scattering from NO, it was illustrative nonetheless. The three sets of cross sections are compared in figure 1, where the differences between them are manifest.

3. Prediction of NO Infrared Emissions

Measurements by Espy *et al.* [8] of the NO(1→0) emission during an aurora are shown in figure 2, as brightness (apparently at zenith, although this is not explicitly stated) as a function of height. The large scatter in the measurements is ascribed by Espy *et al.* to an instrumental problem. They present a fit to the data, using the measurement error as a guide to the weighting given to each data point, but they note that "the fitted curve probably underestimates the intensity at the higher altitudes".

Using the methods described by Campbell and Brunger [1], predictions of these emissions are made for a 3-minute aurora with a 557.7-nm brightness [11] of 40 kR, added on to a background NO density estimated for the solar and auroral conditions for the time of the measurement.

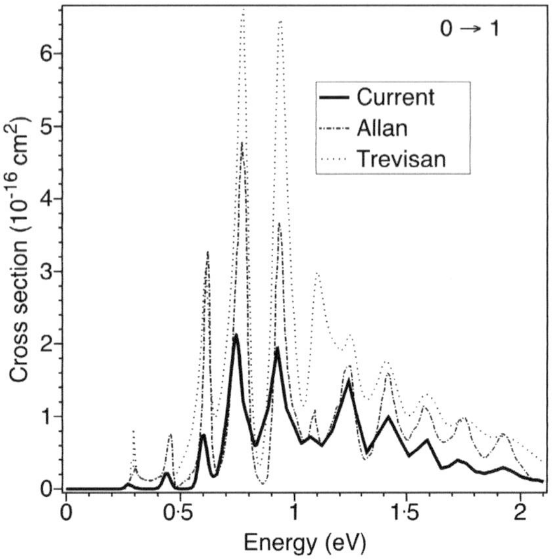

Figure 1. Integral cross sections for electron impact vibrational excitation (0→1) of NO from measurements [3] (———) and theory [6] (- - -), and the 4π-scaled measurements of Allan [6] (– · –).

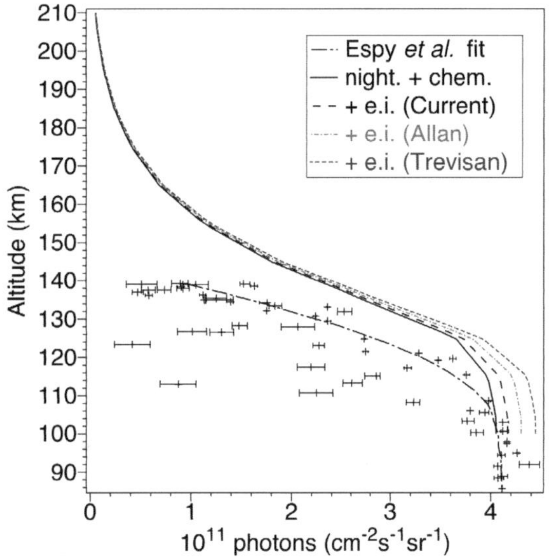

Figure 2. Measurements (|—|—|) of the brightness (looking upwards at the height on the vertical axis) of NO(1→0) emission and a fit (——— –) to them [8]. Predictions for nightglow plus chemiluminescent reactions only (———) and nightglow plus chemiluminescence plus electron impact for the "current" cross sections [3] (– – – –), the 4π-scaled values of Allan [6] (– · –) and the theoretical calculations of Trevisan *et al.* [6] (- - -) are shown.

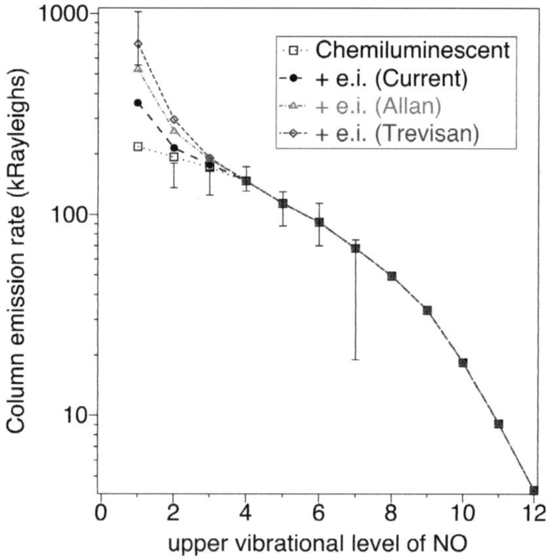

Figure 3. Measured [8] NO fundamental infrared emissions (I) as a function of the vibrational level of NO. Also shown are our predictions for chemiluminescent reactions only (··□··), and for chemiluminescent reactions plus electron impact calculated for the current cross sections (● – – ●), the 4π-scaled values of Allan (△–·–△) and the theoretical calculations of Trevisan *et al.* (◇ - - - ◇).

The predictions are made for the three cross-section sets *i.e.* the "current" case [3], the 4π-scaled values of Allan [6] and the theoretical values [6]. At 100 km the predicted brightness is consistent with the measurements for all three cross-section sets. As the 1→0 emission is dominated by the nightglow, which is proportional to the NO density and therefore has an uncertainty of $\sim$ 10% [2], no particular significance can be attached to the fact that one set gives better agreement than another. At higher altitudes the predicted values are larger than the measurements, but this is consistent with the likely "underestimation" referred to by Espy *et al.*

In figure 3 the predictions using the three cross-section sets are compared with values of the column emission rate (*i.e.* brightness looking towards zenith) for the fundamental ($\nu' \rightarrow \nu' - 1$) auroral emissions from NO calculated by Espy *et al.* from their measurements. The 557.7-nm brightness of 40 kR used in the calculations is much higher than the 5 kR specified by Espy *et al.* for the time of the measurements, but is chosen to match the predicted emission due to chemiluminescent reactions to the observations for $\nu' > 3$.

Espy *et al.* ascribed the peak in the auroral component for $\nu' = 1$ to the difficulty of separating it from the nightglow. However, as shown earlier [1], the inclusion of electron impact, based on the "current" cross sections [3] accounts for part of that peak for $\nu' = 1$. The predicted peak at $\nu' = 1$ is larger when the 4π-scaled values of Allan [6] are used. For the recent theoretical cross sections [6] the electron impact contribution is larger again and brings the prediction into good agreement with the measurements. However, the enhancement of the 2→1 emissions using the recent cross sections increases the discrepancy with the measured values. A possible explanation is that the new electron impact cross sections for 0→2 excitation are too large and that the "current" values are better. Another possibility is that the separation of the different emissions by Espy *et al.* was based on the assumption of a rotational temperature of 300 K for the chemiluminescent emissions, whereas the rotational temperature is likely to be different for electron impact.

4. Conclusions

A previous study involving predictions of auroral infrared emissions from nitric oxide has been repeated for more recent experimental and theoretical electron impact integral cross sections for vibrational excitation of NO. The more recent cross sections, particularly the theoretical, lead to predictions of a larger contribution due to electron impact. For the $1 \rightarrow 0$ emission the more recent experimental cross sections account for most of the previously unexplained spectral peak for this line, while the theoretical cross sections bring the predictions for $1 \rightarrow 0$ into agreement with the measurements. However, the discrepancy for the $2 \rightarrow 1$ emission is increased when using the more recent cross sections. Thus it would be advantageous to resolve the differences between different theoretical and experimental determinations of the electron impact vibrational cross sections in order to perform more accurate modeling of NO infrared emissions in the upper atmosphere.

Acknowledgments

This work was supported by the Australian Research Council. One of us (LC) thanks the Department of Education, Science and Training (DEST), through its International Science Linkages programme, for partial financial support to visit the University of Fribourg.

References

[1] Campbell L and Brunger M J 2007 *Geophys. Res. Lett.* **34** L22102, doi:10.1029/2007GL031743
[2] Campbell L, Brunger M J and Cartwright D C 2007 *J. Geophys. Res.* **112** A08303, doi:10.1029/2007JA012337
[3] Campbell L, Brunger M J, Petrovic Z Lj, Jelisavcic M, Panajotovic R and Buckman S J 2004 *Geo. Res. Lett.* **31** L10103, doi:10.1029/2003GL01915
[4] Josić L, Wróblewski T, Petrović Z Lj, Mechlińska-Drewko J and Karwasz G P 2001 *Chem. Phys. Lett.* **350** 31824
[5] Jelisavcic M, Panajotovic R and Buckman S J 2003 *Phys. Rev. Lett.* **90** 203201–4

[6] Trevisan C S, Houfek K, Zhang Z, Orel A E, McCurdy C W and Rescigno T N 2005 *Phys. Rev.* A **71**, 052714, doi: 10.1103/PhysRevA.71.052714
[7] Allan M 2005 *J. Phys. B: At. Mol. Opt. Phys.* **38** 603-14
[8] Espy P J, Harris C R, Steed A J, Ulwick J C, Haycock H and Straka R 1988 *Planet. Space Sci.* **36** 543–51
[9] Barth C A 1992 *Planet. Space Sci.* **40** 315–36
[10] Caledonia G E and Kennealy J P 1982 *Planet. Space Sci.* **30** 1043–56
[11] Campbell L, Cartwright D C, Brunger M J and Teubner P J O 2006 *J. Geophys. Res.* **111** A09317, doi:10.1029/2005JA011292

Journal of Physics: Conference Series **115** (2008) 012004 doi:10.1088/1742-6596/115/1/012004

Application of the BEf-scaling approach to electron impact excitation of diople-allowed electronic states in molecules

M J Brunger[1], P A Thorn[1], L Campbell[1], H Kato[2], H Kawahara[2], M Hoshino[2], H Tanaka[2] and Y-K Kim[3]*

[1] ARC Centre for Antimatter-Matter Studies, SoCPES, Flinders University, GPO Box 2100, Adelaide, SA 5001, Australia

[2] Physics Department, Sophia University, Chiyoda-ku, Tokyo 102–8554, Japan

[3] National Institute of Standards and Technology, Gaithersburg, MD 20899–8422, USA

E-mail: michael.brunger@flinders.edu.au

Abstract. We consider the efficacy of the BEf-scaling approach, in calculating reliable integral cross sections for electron impact excitation of dipole-allowed electronic states in molecules. We will demonstrate, using specific examples in H_2, CO and H_2O, that this relatively simple procedure can generate quite accurate integral cross sections which compare well with available experimental data. Finally, we will briefly consider the ramifications of this to atmospheric and other types of modelling studies.

*deceased
PACS#: 34.80.Gs

1. Introduction

Accurate theoretical and experimental integral cross sections for electron impact excitations are available for a number of light and heavy neutral atoms. Through comparisons to such data, the BEf-scaling [1] approach was found to convert plane-wave Born (hereafter referred to simply as Born) integral cross sections (ICSs) to reliable cross sections comparable to accurate excitation cross sections at all incident electron energies above threshold. This scaling method provided not only a simple way to calculate excitation cross sections of many atoms, but also to accurately estimate the excitation-autoionisation contributions in many open-shell atoms so that reliable total ionisation cross sections could also be obtained for such atoms [2,3]. If the BEf-scaling method was found to also be applicable to molecules, then it would potentially lead to valid excitation and total ionisation cross sections for many molecules, for which more rigorous collision theories might not be currently used because, for instance, of their complex molecular structure. Indeed in view of the paucity of viable theoretical methods to calculate accurate excitation cross sections directly from the Schrödinger equation, even for many simple diatomic molecules, the scaling method discussed later might provide a welcome relief for those who need such cross sections in their applications, *e.g.*, astrophysics, atmospheric physics, materials processing, semiconductor fabrication and combustion modelling.

Validating a scaling method for Born cross sections of molecules requires two initial ingredients:

(i) the Born integral cross sections themselves (or generalised oscillator strengths from which the ICS can be derived),

(ii) reliable experimental or theoretical optical oscillator strengths ($f_\circ$). Note that this requirement limits the BEf-scaling approach to electric-dipole allowed transitions.

In the first case (i) Born cross sections for simple molecules have been calculated since computers became powerful enough to perform such calculations (late 1960s and onwards), but it is now rare to find a paper on Born cross sections for electronic excitations of molecules. This therefore represents a possible impediment in the implementation of the BEf-scaling approach. In the second case (ii) Berkowitz [4] has tabulated "f sum rule consistent" optical oscillator strengths for many molecules, including H_2, CO and H_2O. We now test the success or failure of the BEf-scaling procedure by considering its results for some representative molecular-electronic transitions, and respectively comparing those results against relevant reliable experimental ICSs. Specifically, we will do this for the $B^1\Sigma_u^+ \leftarrow X^1\Sigma_g^+$ transition in H_2, the $A^1\Pi \leftarrow X^1\Sigma^+$ transition in CO and the $\tilde{A}^1B_1 \leftarrow \tilde{X}^1A_1$ transition in H_2O. Note that this comparison will be made in section 3 of this paper. Prior to doing so, however, a brief description of the BEf-scaling approach will be outlined in section 2, with conclusions finally being drawn in section 4.

2. Theory

Scaled (plane wave) Born cross sections for dipole-allowed excitations – f-scaling and BE-scaling – are described in detail in Kim [5], so only the more important points are mentioned here. Note that the Born cross sections are not only subject to the approximations in the collision theory part, but also depend on the accuracy of the wavefunctions used for the initial and final states of the target molecule.

The f-scaled Born cross section, σ_f, is given by:

$$\sigma_f(E_\circ) = \frac{f_{\text{accur}}}{f_{\text{Born}}}\sigma_{\text{Born}}(E_\circ) \tag{1}$$

where $E_\circ$ is the incident electron energy, f_{accur} is an accurate dipole f value from experiments (*e.g.* dipole (e,e) or photoabsorption) or from accurate wavefunctions, and f_{Born} is the dipole f value from the same wavefunctions used to calculate the unscaled Born cross section, σ_{Born}. The f-scaling process has the effect of replacing the wavefunctions used for σ_{Born} with accurate wavefunctions.

The BE-scaled Born cross section, σ_{BE}, is given by:

$$\sigma_{BE}(E_\circ) = \frac{E_\circ}{E_\circ + B + E}\sigma_{\text{Born}}(E_\circ) \tag{2}$$

where B is the binding energy of the electron being excited, and E is the excitation energy. The BE-scaling corrects the deficiency of the Born approximation at low $E_\circ$, without losing its well-known validity at high $E_\circ$.

If an unscaled σ_{Born} is obtained from poor quality wavefunctions while an accurate f value is known, then both f-scaling and BE-scaling can be applied to obtain a BEf-scaled Born cross section, σ_{BEf}:

$$\sigma_{BEf}(E_\circ) = \frac{f_{\text{accur}}E_\circ}{f_{\text{Born}}(E_\circ + B + E)}\sigma_{\text{Born}}(E_\circ). \tag{3}$$

We note that it is the $\sigma_{BEf}(E_\circ)$ integral cross sections that we compare against corresponding experimental values in the next section of this paper.

For the $\sigma_{BEf}(E_\circ)$ presented in section 3, in H_2 we used the $B^1\Sigma_u^+$ GOSs published by Liu and Hagstrom [6]. Note that their GOSs were a factor of 2 higher than the standard definition

Journal of Physics: Conference Series **115** (2008) 012004 doi:10.1088/1742-6596/115/1/012004

of the GOS [5], because they used excitation energies in Hartrees rather than Rydbergs as the standard definition requires. This was corrected for in the calculation of $\sigma_{\rm Born}(E_\circ)$ by Kim [5]. The theoretical f value deduced from the GOS by Liu and Hagstrom [6] was 0.321237, while the sum of the f values adopted by Berkowitz [4] was 0.310594. These values of f ($f_{\rm Born}$ and $f_{\rm accur}$ respectively) were used in the f-scaling process for $\mathrm{B}^1\Sigma_\mathrm{u}^+$ in H_2. In the case of CO for $\mathrm{A}^1\Pi(\nu')$ we chose the theoretical GOSs of Chantranupong *et al.* [7] to generate $\sigma_{\rm Born}(E_\circ)$, while the accurate optical oscillator strengths found in Berkowitz [4] were employed in the f-scaling. Finally, in the case of water for $\tilde{\mathrm{A}}^1\mathrm{B}_1$ excitation, two sets of theoretical GOSs are available in the literature [8,9]. Bhanuprakash *et al.* [8] calculated their GOS with multiconfiguration wavefunctions, while Durante *et al.*[9] used wavefunctions from the Random Phase approximation. The f value from the GOS by Bhanuprakash *et al.* is 0.054 and the f value from the GOS by Durante *et al.* is 0.046, which is in excellent agreement with that recommended by Berkowitz [4]. Hence, in water we chose the theoretical GOS from Durante *et al.* to generate $\sigma_{\rm Born}$ for the $\tilde{\mathrm{A}}^1\mathrm{B}_1$ electronic state. BEf-scaling was then again applied to obtain ICSs that are in principle valid at low and high $E_\circ$ (see next section).

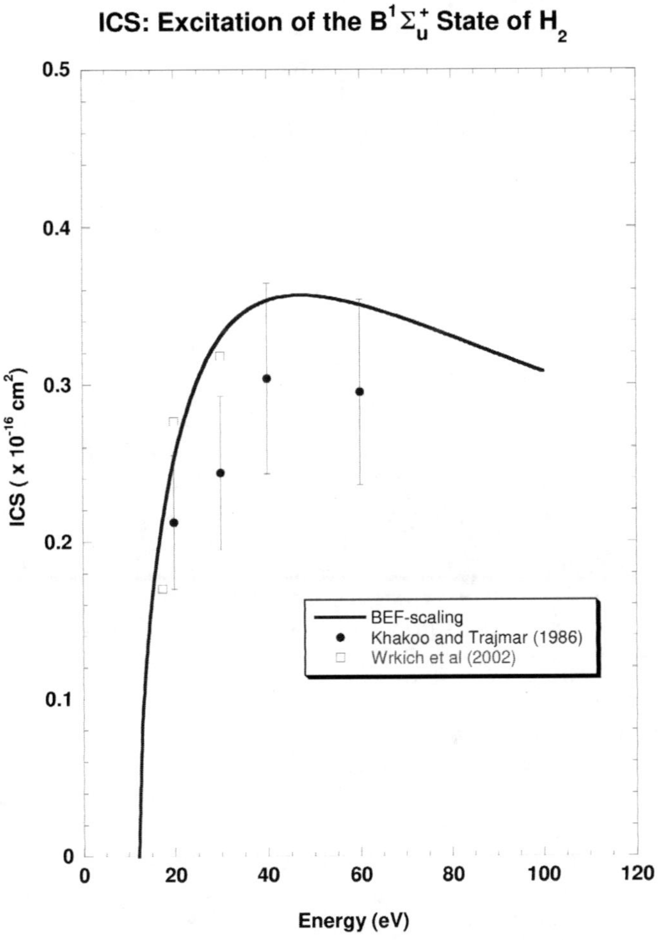

Figure 1. Integral cross sections ($\times 10^{-16}$cm^2) for electron impact excitation of the $\mathrm{B}^1\Sigma_\mathrm{u}^+$ state of H_2. The BEf-scaling result (——) is compared to experimental data from Khakoo and Trajmar [10] (•) and Wrkich *et al.* (□).

3. Results and Discussion

Considering first the electron impact excitation of the $B^1\Sigma_u^+$ electronic state of H_2, we illustrate in Figure 1 the BEf-scaling ICS results of Kim [5] and the available ICS experimental data from Khakoo and Trajmar [10] and Wrkich *et al.* [11]. Even though the energy range of the measured data is somewhat limited and the number of data points relatively few, it is apparent from Figure 1 that there is, to within the stated measurement uncertainties, quite good agreement between the BEf-scaling result and the data [10,11]. Note that this latter statement holds for both the shape and magnitude of the ICS, and therefore provides some nice initial evidence for the utility of the BEf-scaling approach as applied to molecules.

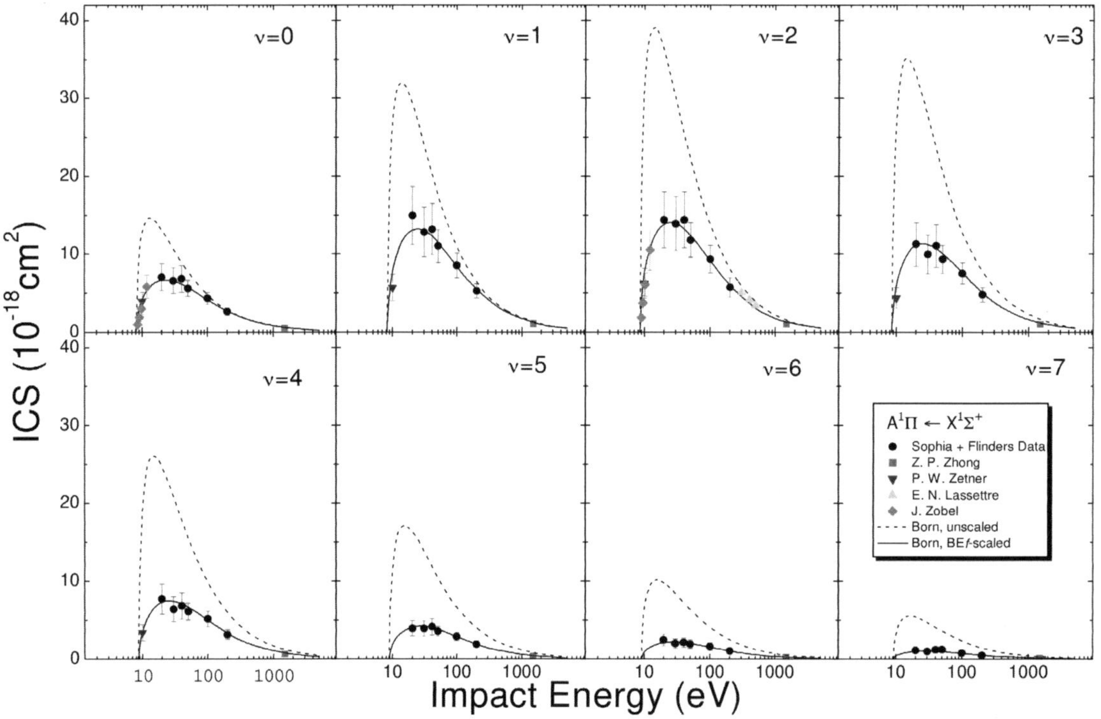

Figure 2. Integral cross sections ($\times 10^{-18}$cm^2) for electron impact excitation of the $A^1\Pi(\nu')$ states of CO. The BEf-scaling results [16] (——), unscaled Born results [16] (- - - -) and experimental data from Zobel *et al.* (♦), Zetner *et al.* [14] (▼), Lassettre and Skerbele [12] (▲), Zhong *et al.* [15] (■) and Kato *et al.* [16] (•) are shown.

Let us now progress from a homonuclear diatomic (H_2) to a weakly polar heteronuclear diatomic (CO). In this case we also consider each of the vibrationally-resolved ($\nu' = 0 - 7$) sub-levels of the excited electronic $A^1\Pi$ state in CO. Here there is also more experimental data available [12–16] for comparison and these experimental ICSs range in energy from near-threshold up to 1500 eV. This comparison is made in Figure 2, where we find an excellent level of agreement between the BEf-scaling result and the measured data [12–16] , over the common energy range, for each $A^1\Pi(\nu')$ vibrational sub-level. Note that the expected inadequacy of the unscaled-Born cross sections is also shown in this figure. The level of agreement seen in Figure 2 between the BEf-scaled result [16] and the data [12–16] is almost unprecedented in our experience [17], so that Figure 2 provides strong evidence for the applicability of the BEf-scaling approach in calculating ICSs for dipole -allowed electronic state transitions in molecules. Unfortunately this observation breaks down when resonance effects are present in the scattering process, as was recently demonstrated by Kawahara *et al.* [18] for the $E^1\Pi$ electronic state in CO. Here they found that the BEf-scaling ICS result was in good agreement with the data [13,15,18] away from the resonance, but only in marginal agreement in the region where the resonance was strongest (near-threshold in this case). This work by Kawahara *et al.* [18] illustrates nicely why the BEf-scaling approach could never hope to supplant *ab initio* scattering calculations, it simply cannot provide a mechanistic description of the scattering process. Nevertheless where such calculations are impractical or unavailable at this time, we believe the BEf-scaling procedure does offer a very useful alternative.

Finally, in Figure 3, we extend our discussion to the polar triatomic molecule H_2O. Specifically, here we look at the electron impact excitation of its $\tilde{A}^1B_1$ electronic state. In this case we compare the BEf-scaling ICS [19] to measured data from Thorn *et al.* [19] and Lassettre and colleagues [20,21]. As before, it is clear from Figure 3 that, to within the uncertainties in the measurements, there is very good agreement between the BEf-scaling result and the ICS data over the common energy range. Species such as water currently represent a formidable challenge to state-of-the-art scattering calculations, so that it is for molecules like H_2O that we see the BEf-scaling approach finding a really useful niche.

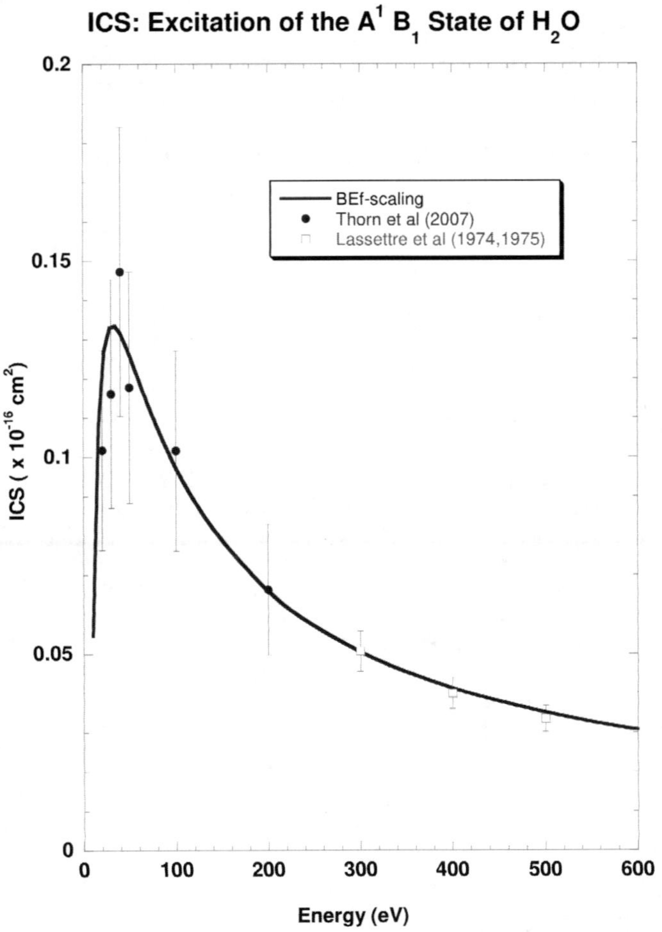

Figure 3. Integral cross sections ($\times10^{-16}$cm^2) for electron impact excitation of the $\tilde{A}^1B_1$ electronic state of H_2O. The BEf-scaling result (——) is compared to experimental data from Thorn *et al.* [19] (•) and Lassettre and colleagues [20,21] (□).

4. Conclusions

This short paper has examined the validity of the BEf-scaling approach to calculating integral cross sections for dipole-allowed excitations in diatomic and polyatomic molecules. While the BEf-scaling approach does not particularly enlighten us as to the fundamental dynamics behind the scattering mechanism in a given system, it does nonetheless appear to predict accurate integral cross sections. This is especially true away from energy regions where resonance phenomena are important. Given that the BEf-scaling technique is relatively simple compared to state-of-the-art *ab initio* theories, we believe it will be valuable for modellers in academia and industry. These modellers require a means of generating reliable cross sections, in order to study the phenomena they are interested in, and we believe the BEf-scaling approach provides precisely that.

Acknowledgments

The work in Australia was supported by the Australian Research Council, while that at Sophia University was supported by the Ministry of Education, Sport, Culture and Technology. The work at NIST was supported in part by the US Department of Energy, Office of Fusion Energy Sciences.

References

[1] Kim Y-K 2001 *Phys. Rev.* A **64**, 032713
[2] Kim Y-K and Stone P M 2001 *Phys. Rev.* A **64** 052707
[3] Kim Y-K and Desclaux J P 2002 *Phys. Rev.* A **66** 012708
[4] Berkowitz J 2002 *Atomic and Molecular Photoabsorption. Absolute Total Cross Sections* (San Diego: Academic Press)
[5] Kim Y-K 2007 *J. Chem. Phys.* **126** 064305
[6] Lin J W and Hagstrom S 1993 *Phys. Rev.* A **48** 166
[7] Chantranupong L, Hirsch G, Bhanuprakash K, Buenker R J, Kimura M and Dillon M A 1992 *Chem. Phys.* **164** 183
[8] Bhanuprakash K, Chandra P, Chabalowski C and Buenker R J 1989 *Chem. Phys.* **138** 215
[9] Durante N, Lamanna U T, Arrighini G P and Buidotti C 1995 *Theor. Chim. Acta* **90** 115
[10] Khakoo M A and Trajmar S 1986 *Phys. Rev.* A **34** 146
[11] Wrkich J, Mathews D, Kanik I, Trajmar S and Khakoo M A 2002 *J. Phys.* B **35** 4695
[12] Lassettre E N and Skerbele A 1971 *J. Chem. Phys.* **54** 1597
[13] Zobel J, Mayer U, Jung K and Ehrhardt H 1996 *J. Phys.* B **29** 813
[14] Zetner P, Kanik I and Trajmar S 1998 *J. Phys.* B **31** 2395
[15] Zhong Z P, Feng R F, Xu K Z, Zhu L F, Zhang J, Ji Q and Shi Q C 1997 *Phys Rev.* A **55** 1799
[16] Kato H, Kawahara H, Hoshino M, Tanaka H, Brunger M J and Kim Y-K 2007 *J. Chem Phys.* **126** 064307
[17] Brunger M J and Buckman S J 2002 *Phys. Repts.* **357** 215
[18] Kawahara H, Kato H, Hoshino M, Tanaka H and Brunger M J, 2007 *Proc. of 25th. ICPEAC (Freiburg, Germany)*
[19] Thorn P A, Brunger M J, Teubner P J O, Diakomichalis N, Maddern T, Boloriazadeh M A, Newell W R, Kato H, Hoshino M, Tanaka H, Cho H, and Kim Y-K 2007 *J. Chem. Phys.* **126** 064306
[20] Lassettre E N and Skerbele A 1974 *J. Chem. Phys.* **60** 2464
[21] Klump K N and Lassettre E N, 1975 *Can. J. Phys.* **53** 1825

Journal of Physics: Conference Series **115** (2008) 012005 doi:10.1088/1742-6596/115/1/012005

Low energy elastic electron scattering from polyatomic targets.

M. A. Khakoo

Department of Physics, California State University, Fullerton, CA 92831, USA.

Abstract: New differential cross-section measurements for elastic electron scattering from ethylene (C_2H_4), three primary alcohols, methanol (CH_3OH), ethanol (C_2H_5OH) and propanol (C_3H_7OH) are reported. The measurements are obtained using the relative flow method with a thin aperture as the collimating target gas source. The relative flow method is applied without the molecular diameters restriction imposed by the relative flow pressure condition on helium (the calibrating gas) and the unknown gases (the primary alcohols). The experimental data were taken at incident electron energies of 1eV, 2eV, 5eV, 10eV, 15eV, 20eV, 30eV, 50eV and 100eV, but only a brief survey of these results will be made here. The experimental results are compared to theoretical differential cross-sections are obtained by using the variational multi-channel Schwinger method. Initial comparisons between theory and experiment show that present theory is well-able to model low electron scattering from these polyatomic targets.

Pacs #s: 34.80.Bm

1. Introduction.

Accurate differential elastic scattering cross-sections for low energy electrons from atomic and molecular gases have provided very useful tests of electron scattering theory, and has resulted in measurement/modeling of other differential scattering processes such as excitation, ionization and electron attachment [1]. These processes play important roles in a range of processes in plasmas [2, 3], astrophysics [4] and in DNA break-up [5]. Elastic electron scattering, a dominant process in the transport of low energy electrons through gaseous media and condensed matter, is important in our understanding of the transport of electrons in various media such as organic tissue made of biomolecules [6], planetary atmospheres [4] and interstellar media, lasers and fusion plasmas [7].

Recently, we developed a novel version of the relative flow method using an aperture gas collimating source, to replace our conventional tube collimating source [8]. The aim was to be able to apply the relative flow method *without* the restriction imposed by the requirement that the pressures behind the source be in the ratio of the gas kinetic molecular diameter squared. This is because the aperture source's gas profile is not dependent on the source pressure over a large pressure range up to when the mean-free path of the gas behind the source becomes comparable to the thickness of the aperture. This innovation is therefore useful for large molecules where the values of these diameters are not available or not accurately known. A series of recent measurements in Tetrahydrofuran [9,10,11] have used a molecular diameter from a very old source [12]. Whereas these measurements show good qualitative agreements (shapes), there exist some quantitative disagreements (magnitudes) and the situation could be improved, if the molecular diameter requirements in these experiments were eliminated. A recent important synopsis of these measurements which includes further work using an aperture-like source by Allen [13], shows that there exists a significant dependence on the pressure ratios used which affect the cross-section ratios.

Presently, there are no measurements of elastic electron scattering (differential) cross-sections from most polyatomics molecules since the gas kinetic molecular diameters of these polyatomics are unavailable or unreliable. In particular, for the alcohols, or molecules with permanent dipole moments, one expects the gas kinetic molecular diameter to deviate significantly from that of a hard-sphere type of molecule, and make these sticky targets more difficult to work with in the conventional relative flow type of experiment.

In this paper report the extension of our recent implementation of the relative flow method, using an aperture source [8], to alcohols. The proposed method is a simple modification of the conventional relative flow method in which a collimating tube source is replaced by an aperture source. The results are supported by new large theoretical calculations using the variational multi-channel Schwinger method with polarization effects included [14].

2. Method

The most successful method that has been used for measuring differential elastic electron scattering from relatively light atomic and molecular targets has been the relative flow method, which was first formulated by Srivastava et al. [15] and has been applied extensively in the last 30 years to determine elastic electron scattering DCSs for many (mostly light) targets. The reader is referred to the extensive review of Brunger and Buckman [1] for the most recent summary of this work. The experimental apparatus (spectrometer, vacuum chamber, control equipment) has been detailed in previous papers, e.g. Khakoo *et al.* [8] . The target aperture of 0.3mm diameter was located on a thin disc of 0.025mm thick brass shim stock located flush at the end of a 6.35mm o.d. and 4.3mm i.d. brass tube [8]. The tube was incorporated into a moveable source [16] arrangement.

We followed the recipe given in [17] to calibrate our relative flow rate (RFR) vs. the pressure behind the source P_s. A second-order polynomial in the input gas pressure behind the source P_s served adequately to express relative flow rate as a function of P_s in the form $RFR=aP_s(1+\varepsilon P_s)$. The coefficients a and ε are related to the mass and the molecular diameters of the gases [18,19]. These coefficients are given in table 1 for all gases used in our aperture gas system to date. As can be seen the gas kinetic molecular diameters determined from a and ε values from our flow measurements for the alcohols are large, typifying the long-range dipole forces these molecules must exert on each other as compared to hard-sphere molecules. However, the 'a' term for these molecules deviates markedly by a factor of $\approx\sqrt{2}$ from its molecular mass dependence (last column in table 1; Mass Factor). The reason for this is not certain, but it could be speculated that these molecules form dimers during their flow through the gas system from the liquid phase into the gas phase. This is under discusson at the present time.

Journal of Physics: Conference Series **115** (2008) 012005 doi:10.1088/1742-6596/115/1/012005

We operated our experiment at low pressures, typically <2 Torr for He, <0.2 Torr for CH_3OH, <0.12 for C_2H_5OH and <0.1Torr for C_3H_7OH (n-propanol), which are well below the critical mean-free path (when this equals the aperture thickness of 0.025mm) conditions for our system. The experimental chamber typically remained in the pressure range of 0.7 to 2 x10^{-6} Torr for the range of pressure used in this work and the incident electron current remained stable during operation within 10% at maximum.

Gas	T (C)	P^2	P	a	ε	δ (cm)	M	a*√M	ε/δ^2	Mass Factor
H_2	22	0.2290	0.2181	0.2181	1.0501	2.74E-08	2.016	0.310	1.40E+15	1
He	22	0.1146	0.1551	0.1551	0.7391	2.18E-08	4.002	0.310	1.56E+15	1
He	47	0.1028	0.1611	0.1611	0.6380	2.18E-08	4.002	0.322	1.34E+15	1
He	74	0.1064	0.1615	0.1615	0.6587	2.18E-08	4.002	0.323	1.39E+15	1
N_2	22	0.1261	0.0596	0.0596	2.1145	3.75E-08	28.02	0.316	1.50E+15	1
C_2H_4	22	0.2305	0.0566	0.0566	4.0705	4.95E-08	28.03	0.300	1.66E+15	1
CH_3OH	47	0.2214	0.0387	0.0387	5.7222	6.20E-08	32.04	0.310	1.49E+15	2
C_2H_5OH	47	0.2582	0.0352	0.0352	7.3462	7.02E-08	46.07	0.337	1.49E+15	2
C_3H_7OH	74	0.2328	0.0288	0.0288	8.0953	7.35E-08	60.11	0.315	1.50E+15	2

Table 1. Gas flow parameters for the aperture source for various gases. See text for discussion.

We took measurements of elastic scattering from CH_3OH and C_2H_5OH and C_2H_7OH at E_0 values of 1eV, 2eV, 5eV, 10eV, 15eV, 20eV, 30eV, 50eV and 100eV at scattering angles of 5° to 130°, over a coarser interval (20°) at large scattering angles than at small scattering angles (10°). The data were normalized via the relative flow method to the established He DCS values from [20] and [21].

3. Results:

A brief review of our DCS measurements is reported here. More complete reports on this are currently in progress. The methanol DCSs are compared to theory in figures 1 and for ethanol in figures 2. The Schwinger multi-channel theory both includes (Born dipole-polarization) and excludes (static) dipole-polarization effects and was developed and computed by theoretical groups in Brazil and the USA [14]. Agreement of our experiment theory is qualitatively excellent at all E_0 values for CH_3OH and C_2H_5OH. It is especially good at lower E_0 values where both the dipole-polarization calculations from Brazil and from USA show agreement. Clearly, getting the dipole-polarization correct becomes very important for theory at low E_0 values, e.g. at 5eV or below when compared to only the static potential calculations. At E_0=10eV, 15eV, 20eV and 30eV agreement between theory and experiment is very good.

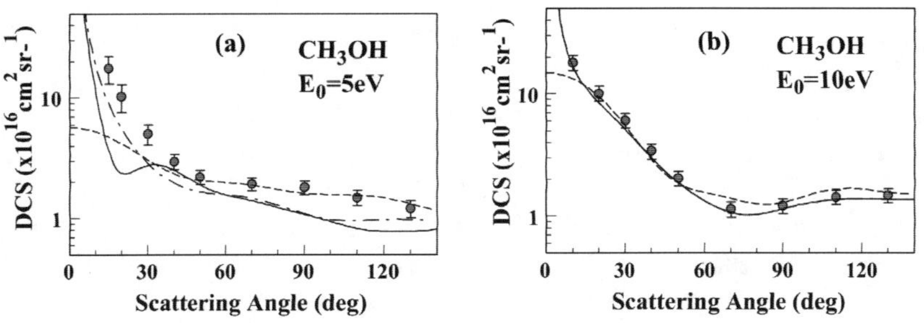

Journal of Physics: Conference Series **115** (2008) 012005 doi:10.1088/1742-6596/115/1/012005

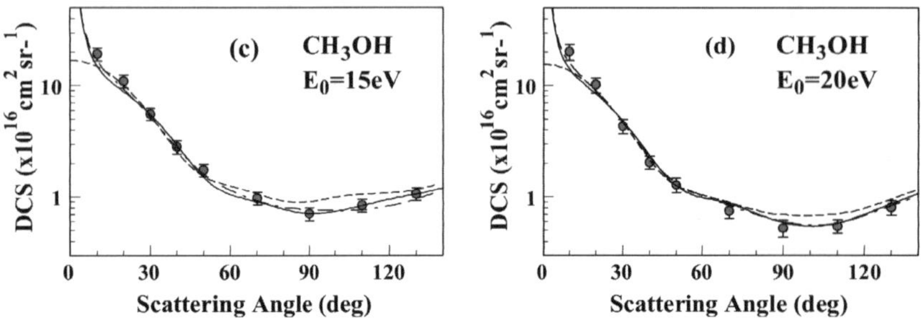

Figure 1 a-d. Selected DCSs for elastic scattering from methanol. • Present experiment; Present theories: — multi-channel Schwinger with polarization (USA), — - — multi-channel Schwinger with polarization (Brazil); ---- multi-channel Schwinger static (USA/Brazil) [14].

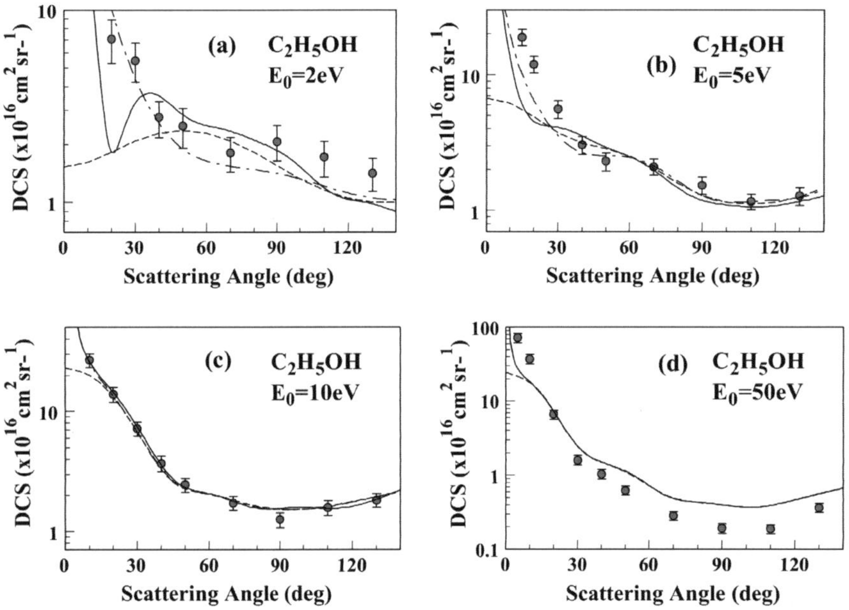

Figure 2 a-d. Selected DCSs for elastic scattering from ethanol. • Present experiment; Present theories: — multi-channel Schwinger with polarization (USA), — - — multi-channel Schwinger with polarization (Brazil); multi-channel Schwinger static (USA/Brazil) [14].

The experiment remains lower than theories for large angles at E_0=100eV and has a steeper rise at small θ values. Even at low energy (e.g. 5eV), polarization in the scattering (steep forward-peaking) is strong, and is also evident to a similar extent for C_2H_5OH. However at higher E_0 values, the differences between theory and experiment are larger for C_2H_5OH than for CH_3OH. The shapes of the DCSs for these 3 alcohols are almost identical as can be observed in figure 3 at E_0=30eV. We note that the small angle DCSs show similarly strong polarization dependence, which is expected as these alcohols have almost equal dipole moments (≈1.6 Debyes). However, at lower energies, we observe significant differences in large angle behavior.

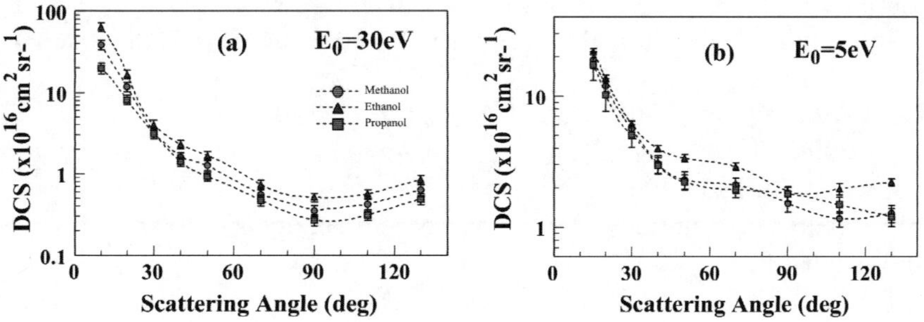

Figures 3. Sample of present DCSs (connected by dashed lines) with error bars for elastic electron scattering from methanol at (a) E_0=30eV and (b) 5eV. Legend: ▲Propanol, • Ethanol and ■ Methanol

4. Conclusions.

In conclusion, the present work shows that theory and experiment have made significant progress towards the determination of DCSs for elastic scattering from large polyatomics, a fact that was not considered possible until recently. Presently we are considering the extension of our measurements to butanol, C_4H_9OH.

5. Acknowledgements

This work was funded by a collaborative grant from the National Science Foundation PHY 0653452 to Caltech and Cal. State Fullerton and by grants from CNPq-Brazil. We acknowledge the theoretical input of Dr. B. Vincent McKoy and Dr. Carl Winstead (Caltech, USA), Dr. Marco A. P. Lima (Campinas, Brazil) and Dr. Marcio Bettega (Curitiba, Brazil), to this report, prior to publication of DCS data. We acknowledge undergraduate students John Muse and Kyle Keane (CSUF), Joe Blumer (Brigham Young U.) and graduate student Helen Silva (Juis de Fora, Brazil, visiting scholar's grant from CNPq-Brazil) for input in these measurements.

6. References.

[1] Brunger M J and Buckman S J 2002 Phys. Rep. **357**, 215.
[2] Whelan C T and Mason N J 2005 *Electron Scattering from Atoms, Molecules Nuclei and Bulk Matter*, Kluwer Academic/Plenum Publishers, New York.
[3] Lister G G, Lawler J E, Lapatovich W P and Godyak V A 2004 Rev. Mod. Phys. **76**, 541.

[4] Strickland D J, Lean J L, Meier R R, Christensen A B, Paxton L J, Morrison D, Craven J D, Walterscheid R L, Judge D L and McMullin D R 2004 Geophys. Res. Letts. **31**, L03801.

[5] Boudaïffa B, Cloutier P, Hunting D, Huels M A and Sanche L 2000 Science **287**, 1658-1660.

[6] Tonzani S and Greene C H 2006 J. Chem. Phys. **125**, 094504.

[7] Stangeby P C 2000 *The Plasma Boundary of Magnetic Fusion Devices,* Institute of Physics Publishing, Philadelphia.

[8] Khakoo M A, Keane K, Campbell C, Guzman N and Hazlett K, 2007 J. Phys. B: At. Mol. Opt. Phys. 40**,** 3601

[9] Milosavljevi´c A R, Giuliani A, ˇSevi´c D, Hubin-Franskin M J and Marinkovi´c B P 2005 Eur. Phys. J. D **35** 411.

[10] Dampc M, Milosavljevi´c A R, Linert I, Marinkovi´c B P and Zubek M 2007 Phys. Rev. A **75** 042710.

[11] Colyer C J, Vizcaino V, Sullivan J P, Brunger M J and Buckman S J 2007 New J. Phys. **9** 41.

[12] *Landholt-Bornstein, Zahlenwerte und Functionen, Bd I, Atom und Molekularphysik, 1, Atome und Ionen*, edited by A. Eucken (Springer-Verlag, Berlin, 1950).

[13] M. Allan, J. Phys. B: At. Mol. Opt. Phys. (2007); private communication; in review.

[14] Analytic Multi-channel Schwinger, with polarization effects, Winstead C, McKoy B V, Lima M A P and Bettega M, Private communication.

[15] Srivastava S K, Chutjian A and Trajmar S 1975 J. Chem Phys. **63**, 2659.

[16] Hughes M, James K E, Jr. J. G. Childers J G and Khakoo M A 2003 Meas. Sci. and Tech. **14**, 841.

[17] Khakoo M A, Jayaweera T, Wang S and Trajmar S 1993 J. Phys. B: At Mol.Opt. Phys. **26,** 4.

[18] Sagara T and Boesten L 1998 J. Phys. B: At. Mol. Opt. Phys. **31**, 3455.

[19] Rugamas F, Roundy D, Mikaelian G, Vitug G, Rudner M, Shih J, Smith D, Segura J and Khakoo M A 2000 Measurement Science and Technology **11**, 1750.

[20] Nesbet R K 1979 Phys. Rev. A **20**, 58.

[21] Register D F, Trajmar S and Srivastava S K 1980 Phys. Rev. A **21**, 1134.

Investigation of dissociative electron attachment to water using ion momentum imaging

N Bhargava Ram, Vaibhav S Prabhudesai[1], P. Rawat, G. Aravind and E Krishnakumar[2]

Tata Institute of Fundamental Research, Colaba, Mumbai 400005 India

Email: ekkumar@tifr.res.in

Abstract. The dynamics of the dissociative electron attachment (DEA) process to water (H_2O) and heavy water (D_2O) are investigated using the newly developed Velocity Map Imaging for low energy electron collisions. Where as the kinetic energy and angular distribution of the negative ion from water at the 6.5 eV resonance is found to be in excellent agreement with previous results, entirely different angular distributions are obtained at the two resonances at higher energies. We identify the 8.5 eV resonance as due to a $^2B_1$ state in contrast to what has been believed till now based on experimental and theoretical studies. Large anisotropy in the angular distribution of H^- and O^- ions is observed at the third resonance indicating interesting dynamics involved.

1. Introduction

Low energy electron collisions on water have been subjected to a great deal of investigations in recent times. These investigations are driven by the fact that water is the most basic molecule of life, is present in wide variety of astrophysical bodies, its relevance to storage and disposal of radioactive waste and it is one of the smallest triatomic systems. Dissociative electron attachment (DEA) is one of the dominant processes in low energy electron collisions and also the most interesting from the molecular dynamics point of view. However, modeling this process even for simple diatomic molecules still remains a very challenging problem. Despite this complexity, dedicated efforts to calculate the DEA process in water in detail have borne fruits [1] and have been able to reproduce the experimental results on absolute cross sections with fair accuracy [2]. Along with the cross sections, the kinetic energies and angular distributions of the fragment ions provide the most detailed input on the dynamics of the DEA process. There are two existing set of measurements on the kinetic energies and angular distributions [3, 4]. However, considering the availability of accurate theoretical calculations and that the new accurate experimental technique of ion momentum imaging is available with us, we investigated the DEA process in water for the kinetic energies and angular distributions of the fragment ions. These measurements are also provoked by our recent observation of functional group dependent DEA in organic molecules leading to site selective fragmentation [5]. We felt it necessary to make detailed measurements of the DEA dynamics in water, ammonia and methane in order to probe the depth to which the characteristics of the DEA process in these molecules are

[1] Present address: Dept. of Phys. of Complex Systems, Weizmann Institute of Science, Rehovot 76100, Israel
[2] Author to whom any correspondence may be addressed

retained in molecules containing O-H, N-H and C-H bonds. Thus the present measurements on water are part of an ongoing set of measurements on a broader scale.

2. Results and Discussion

The DEA process in water is known to give three resonances in the 6 to 14 eV energy range [2]. H^- ions are the dominant species in this process with peaks at 6.5 eV, 8.5 eV and 11.8 eV respectively. The O^- intensities are smaller by about a factor 100 and 10 respectively at the first two resonances and larger by a factor of 2 at the third resonance [2]. The OH^- cross sections are known to be much smaller than that of O^-[6]. The momentum images of H^- and O^- from water was obtained at all the three resonances using the Velocity Map Imaging technique developed in our lab, which has been discussed in recent publications [7, 8].

2.1 First resonance at 6.5 eV

The angular distribution and kinetic energies of H^- ions obtained at the first resonance at 6.5 eV has been discussed already [8]. Our measurements of the angular distribution of H^- ions were in excellent agreement with the previous results [3, 4]. The kinetic energy distribution also was more or less in agreement with the previous results as discussed earlier [8]. The velocity map image of O^- ions at the first resonance was found to be an unresolved small blob, indicating relatively small kinetic energies with almost isotropic angular distribution.

2.2 Second resonance at 8.5 eV

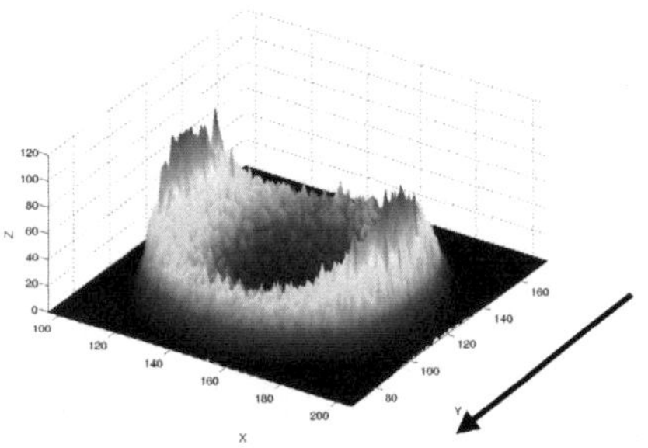

Figure 1. Velocity image of H^- ions at 8.5 eV. The arrow indicates the electron beam direction.

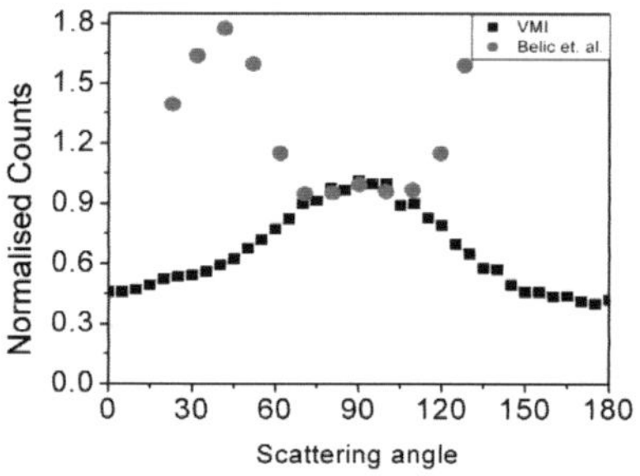

Figure 2. Angular distribution of H^- ions at 8.5 eV normalized at 90°. Squares – present data, circles – Belic *et al.*

Figure 1 shows the velocity map image of H^- ions obtained at the second resonance. The only other measurement of angular distribution of H^- at 8.5 eV is by Belic *et al.*, who used a 127° electrostatic analyzer for measuring angle and energy of the H^- ions. Figure 2 shows the angular distribution data obtained by us along with that by Belic *et al.* It can be seen that there is considerable disagreement between the two. The considered wisdom till now, based on the data by Belic *et al* and various theoretical calculations [1, 9] is that the 8.5 eV resonance is a $^2A_1$ state. However, our angular distribution data show that the resonance is a B_1 state with the electronic configuration $(1b_1)^{-1}$ $(2b_2)^2$. Like in the case of the 6.5 eV resonance, this resonance also gives a distribution of kinetic energies for H^- ions. In this case the energy ranges from 0 to 4 eV, corresponding to the formation of OH radical in various vibrational levels. The observed energies of H^- ions are consistent with the dissociation channel of H^- + OH (X $^2\Pi$) with threshold energy of 4.35 eV.

Journal of Physics: Conference Series **115** (2008) 012006 doi:10.1088/1742-6596/115/1/012006

The velocity image of O^- ions produced at the second resonance was very similar to that observed at the first resonance. The maximum KE observed doesn't exceed 0.4 eV and the dissociation channel is identified to be $O^- + H_2$ (threshold energy 3.6 eV). From the observed O^- energies, it appears that the H_2 will be vibrationally excited. Energetically, the channel $O^- + H + H$ with threshold energy of 8.04 eV is also available at this resonance.

2.3 Third resonance at 11.8 eV

Figure 3 shows the image of the H^- ions formed at the third resonance of 11.8 eV. We see three distinct distributions in the radial direction indicating three distinct kinetic energies of the H^- ions. The two inner ones are clearly visible. Outside of these two there is a relatively weak one seen in light blue shade. For all the three distributions, the intensities in the backward scattering angles are more than those in the forward angles. The kinetic energy distribution of the ions is plotted in figure 4 showing the three distinct distributions, the first one peaking at 1.2 eV, second one at 2.5 eV and third one appearing as a broad structure between 3.5 and 6 eV. The angular distribution of the H^- ions at each of the three energies is given in figure 5 (a, b, c, d) along with the results of Belic *et al.* at 12.0 eV.

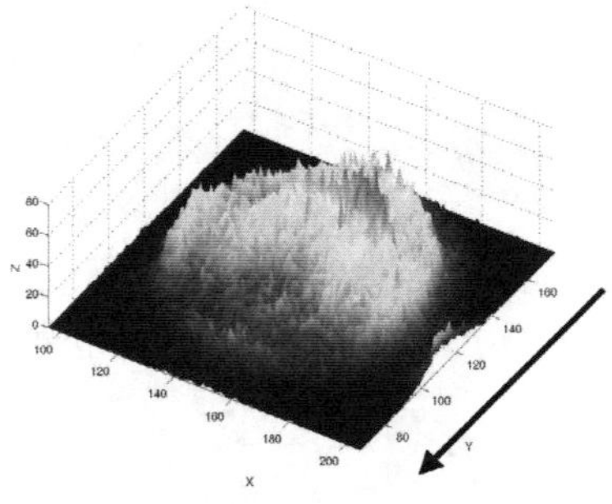

Figure 3. Velocity image of H^- ions at 11.8 eV resonance. The arrow indicates the electron beam direction.

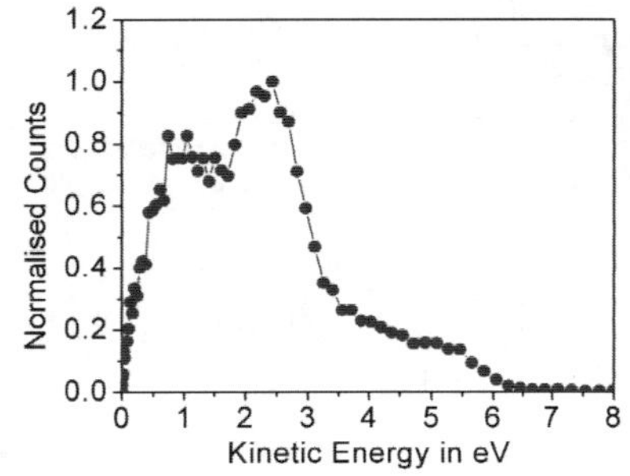

Figure 4. Kinetic energy distribution of H^- ions at 11.8 eV resonance.

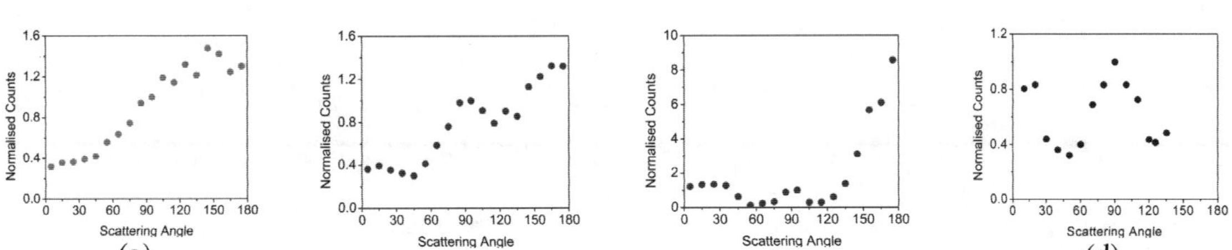

Figure 5. Angular distributions of the H^- ion groups (normalized at 90°) with (a) energy peak at 1.2 eV (b) energy peak at 2.5 eV and (c) energies from 3.5 to 6 eV (d) Angular distribution reported by Belic *et al.* at 12 eV.

The three distinct distributions of H^- energies indicate that there are three dissociation channels for the resonance at 11.8 eV. The innermost distribution of H^- ions (peaking at 1.2 eV) could be identified as the $H^- + H + O$ channel with threshold energy of 8.75 eV. The second group of H^- ions with KE peak at 2.5 eV could be from the $H^- + OH^*$ ($^2\Sigma$) channel with threshold energy of 8.38 eV. The third group of ions with energies ranging from 3.5 eV to about 6 eV appears to arise from the $H^- + OH$ ($X\ ^2\Pi$) channel which has the threshold energy of 4.35 eV. The previous energy and angular distribution measurements of H^- and D^- from DEA to H_2O and D_2O at 12.0 eV by Belic *et al.* report only the H^- +

OH* ($^2\Sigma$) (D^- + OD* ($^2\Sigma$) in the case of D_2O) channel and the KE peak at about 2.7 (2.2 eV for D^-) with sharp rising and falling sides. The kinetic energy measurements by Curtis and Walker [10] for D^- ions from D_2O show two components in the KE distribution arising from the D^- + D + O channel (lower energy component) and the D^- + OD* ($^2\Sigma$) channel (higher energy component). Similar to the case of H_2O discussed above, we have carried out detailed momentum imaging of ions from D_2O. Our results on D^- from D_2O at 11.8 eV show three KE components again, similar to the case of H^- from H_2O.

Figure 6 shows the velocity image of the O^- ions from H_2O at the third resonance at the electron energy of 11.8 eV. Unlike at the first two resonances, at this resonance, O^- ions have a well characterised energy and angular distributions. There ions are scattered mostly in the forward direction and the angular distribution peaks at about 60 degrees as seen in figure 7.

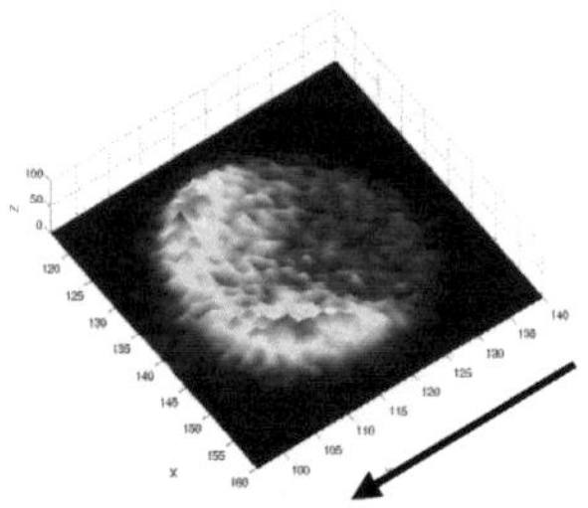

Figure 6. Velocity Map image of O^- from DEA to H_2O at 11.8 eV

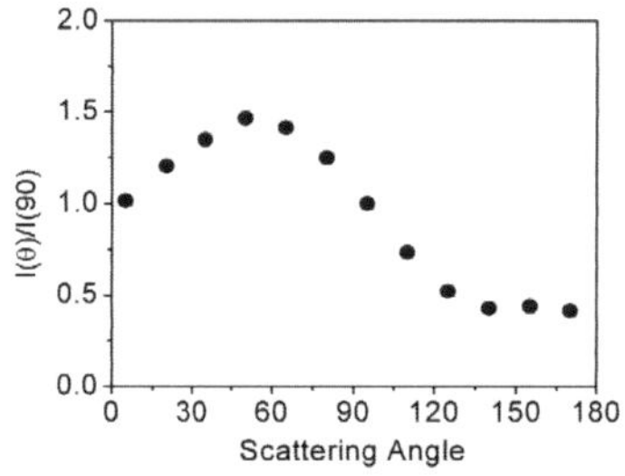

Figure 7. Angular Distribution of O^- ions from DEA to H_2O at 11.8 eV

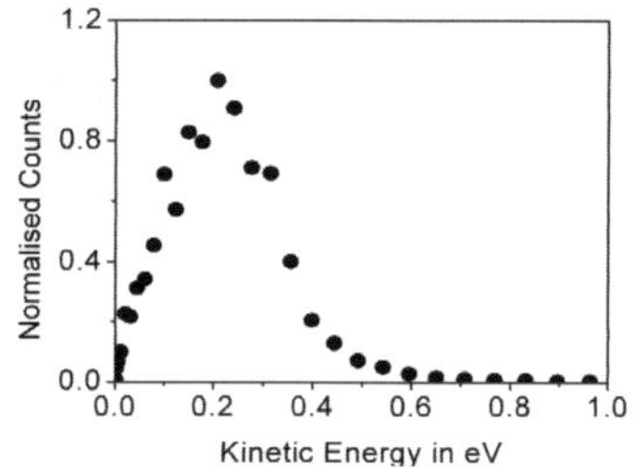

Figure 8. KE distribution of O^- ions from DEA to H_2O at 11.8 eV.

As could be seen from figure 5, the angular distributions of H^- obtained by us for each of the three dissociation channels is different from that reported earlier by Belic *et al.* It is also striking to note that at this resonance, the angular distribution of H^- ions is predominantly in the backward direction, whereas that for O^- ions is in the forward direction (figure 7). We are in the process of analysing these results.

References:

[1] Haxton D J, McCurdy C W and Rescigno T N 2007 *Phys. Rev. A* **75** 012710; Haxton D J, Rescigno T N and McCurdy C W 2007 *Phys. Rev. A* **75** 012711
[2] Rawat P, Prabudesai V S, Aravind G, Rahman M A and Krishnakumar E 2007 *J. Phys. B: At. Mol. Opt. Phys.* **40** 4625
[3] Trajmar S and Hall R I 1974 *J. Phys. B: At. Mol. Phys* **7** L485
[4] Belic D S, Landau M and Hall R I 1981 *J. Phys. B: At. Mol. Phys* **14** 175
[5] Prabhudesai V S, Kelkar A H, Nandi D and Krishnakumar E 2005 *Phys. Rev. Lett.* **95** 143202
[6] Fedor J, Cicman P, Coupier B, Feil S, Winkler M, Głuch K, Husarik J, Jaksch D, Farizon B, Mason N J, Scheier P and M¨ark T D 2006 *J. Phys. B: At. Mol. Opt. Phys.* **39** 3935
[7] Nandi D, Prabhudesai ,Vaibhav S, Chatterjee A and Krishnakumar E 2005 *Rev Sci Instrum* **76** 053107
[8] Bhargava Ram N, Prabhudesai V S, Aravind G, Rawat P and Krishnakumar E, *J. Phys.: Conf. Ser.* **80** 012017
[9] Jungen M Vogt J and Staemmler V 1979 *Chem. Phys.* **37** 49
[10] Curtis M G and Walker I C 1992 *J. Chem. Soc. Faraday Trans.* **88** 2805

Core-excitation processes of O(1s) in CO, CO_2 and OCS molecules by electron impact

F. Arretche, K.T. Mazon, A.S. Falck, A. Marin, H.L. Oliveira, O.A. Pessoa, D. Travessini, S.E. Michelin
Departamento de Física, UFSC, 88040-900 Florianópolis, SC, Brazil
E-mail: farretche@hotmail.com.

M.M. Fujimoto
Departamento de Física, UFPR, 81531-990 Curitiba, PR, Brazil

M.-T. Lee
Departamento de Química, UFSCar, 13565-905, São Carlos, SP, Brazil

Abstract. Distorted-wave approximation is applied to study electron-impact excitation of core electrons in CO_2, CO and OCS. Differential and integral cross sections for the transitions: $X\,^1\Sigma_g^+ \rightarrow ^{1,3}\Pi_u(\,1\sigma_g \rightarrow 2\pi_u)$ in CO_2, $X\,^1\Sigma^+ \rightarrow ^{1,3}\Pi(2\sigma \rightarrow 4\pi)$ in OCS, and $X\,^1\Sigma^+ \rightarrow ^{1,3}\Pi(1\sigma \rightarrow 2\pi)$ in CO are calculated and reported in the (550-1000)-eV incident energy range. Comparison is made among the calculated data for the three targets. The physical origins of the similarity and difference of these data are also discussed. In addition, the generalized oscillator strengths for singlet $1\sigma_g \rightarrow 2\pi_u$ and $1\sigma_u \rightarrow 2\pi_u$ transitions for CO_2 are calculated at 1300 eV. The comparison of these results with the available data in the literature is encouraging. [1]

1. Introduction

Investigations on inner-shell excitation or ionization processes in atoms and molecules, involving the promotion of a strongly linked K-shell electron to an unfilled bound or continuum orbital, are both important and interesting [1]. In general, such internal molecular orbitals are highly localized and their participation in the chemical bonding formation is small. Therefore, they are very similar to the corresponding atomic orbitals. Due to this atomic-like character, the shapes and energies for K-shell electronic excitations do not vary significantly in molecules with common atomic constituent, even in different physical states [2]. This property of the inner-shell molecular orbitals can be used in the development of equipments to characterize both gaseous or solid samples [3].

In the last two decades, the inner-shell excitation and ionization processes have received considerable attention. Experimentally, both high-energy photons (vacuum ultraviolet radiation and X-ray) and electrons are used as excitation sources. Such studies include, for instance, photoexcitation [4, 5, 6], photoemission and photoionization [7, 8], electron-impact excitation

[1] PACS NUMBERS: 32.80.Aa,34.80.Gs

[9, 10, 11, 12, 13] of molecules. On the theoretical side, some solid *ab initio* approaches such as the R-matrix method and the multichannel quantum-defect theory have been applied to study optically stimulated inner-shell excitation of molecules [14, 15, 16]. Although an *ab-initio* multichannel study on electron-impact core-excitation processes is also desirable, computationally couplings between a fast incident electron and a slow outgoing electron is very difficult to deal with. To our knowledge, only studies using the perturbative methods such as the first Born approximation (FBA) [11, 12] and distorted-wave approximation (DWA) [17, 18, 19] can be found in the literature. Despite the simplicity, quite accurate generalized oscillator strengths (GOS) for electron-impact core-excitation of molecules can be provided using these methods. Moreover, the occurrence of shape resonances at incident energies near the excitation thresholds can also be predicted by the DWA.

Recently, Kroin *et al.* [17] reported a theoretical investigation on the excitation of an electron from the carbon 1s orbital, C(1s), to the lowest-lying unfilled π orbital in a series of triatomic molecules containing only one carbon atom, namely CO_2, CS_2, and OCS. They observed that in the non-resonant energy region, the theoretical ratios between the integral cross sections (ICS) for transitions leading to the singlet core-excited states and those leading to triplet excited states, refered here as RI(1:3), are nearly the same for the three targets studied therein. This behaviour seems to confirm the atomic-like property of such orbitals.

On the other hand, for molecules containing two or more equivalent atoms of the same chemical element, the molecular-orbital description of the inner-shell wave functions would require a linear combination of the atomic orbitals. Therefore, they are delocalized and no longer atomic. Recently, we have applied the DWA to study the electron-impact core excitation of two isoelectronic molecules: N_2 [18] and C_2H_2 [19]. In these targets, the linear combination of the 1s orbitals of two carbon (nitrogen) atoms has resulted in the delocalized core molecular orbitals $1\sigma_g$ and $1\sigma_u$ in C_2H_2 (N_2). The different parity of these molecular orbitals leads to distinct behavior in excitation and ionization processes. For instance, the singlet $1\sigma_u \rightarrow 3\pi_g$ transition is dipole-allowed whereas the singlet $1\sigma_g \rightarrow 3\pi_g$ transition is dipole-forbidden. Consequently, the calculated cross sections for the above transitions are expected to be quite different from each other.

In order to better understand the influence of the atomic-like or delocalized character of the K-shell molecular orbitals, in this work we extent the application of the DWA to investigate the electron-impact O(1s) core-excitation processes in the triatomic CO_2 and OCS molecules and in the diatomic CO molecule. The differential cross sections (DCS) and ICS for promotion of an electron from the 1σ orbital in CO, the 2σ orbital in OCS and $1\sigma_g$ orbital in CO_2 to the corresponding lowest unfilled orbital are calculated and compared with each other. The Hartree-Fock (HF) calculations have revealed that the 1σ orbital in CO and the 2σ orbital in OCS are highly localized whereas the $1\sigma_g$ orbital in CO_2 is delocalized. Therefore, the similarity and difference in the calculated cross sections can provide insight of the dynamics of core-shell excitations by electron impact.

The organization of the present article is as follows: in section 2, an outline of the theory used is presented. Some details of the calculations are also provided. Finally in section 3, we compare the calculated data for the three targets and summarize our conclusions.

2. Theory

Details of the basic theory used in this work can be found elsewhere [20, 21, 22, 23]. The electron-impact excitation DCS averaged over molecular orientations is given by:

$$\frac{d\sigma}{d\Omega} = \frac{1}{8\pi^2}\int d\alpha\, sin\beta\, d\beta\, d\gamma\, |f(\hat{k}_i', \hat{k}_f')|^2, \qquad (1)$$

Journal of Physics: Conference Series **115** (2008) 012007 doi:10.1088/1742-6596/115/1/012007

Table 1. Vertical excitation energy (ΔE) and singlet-triplet energy splitting (δ), in eV

Molecule	ΔE for singlet transitions	ΔE for triplet transitions	δ
CO_2	554.018 535.5[a] 535.99[b]	553.337	0.681
OCS	550.960 520.0-560.0[c]	549.746	1.214
CO	550.097	548.486	1.611

[a] Experimental results of Eustatiu *et al.* [12];
[b] Experimental results of Miranda *et al.* [11];
[c] Data from Hitchcock *et al.* [1].

where $\hat{k}'_f(\hat{k}'_i)$ is the linear-momentum direction of the scattered (incident) electron in the Laboratory-Frame (LF), whereas the direction of incident electrons is taken as the LF z-axis. The (α, β, γ) are the Euler angles which define the direction of the molecular principal axes. The Body-Frame (BF) scattering amplitude $f(\hat{k}_i, \hat{k}_f)$ is related to the T-matrix elements by the expression:

$$f(\hat{k}_i, \hat{k}_f) = -2\pi^2 T_{if} \tag{2}$$

and is related to the LF scattering amplitude by usual transformations [24].

Using the DWA, the transition T matrix is given by:

$$T_{if} = \langle \varphi_f \Psi^{(-)}_{k_f} \mid U_{se} \mid \varphi_i \Psi^{(+)}_{k_i} \rangle \tag{3}$$

where φ_i and φ_f are the initial and final target wave functions, respectively. These wave functions are Slater determinants with appropriate symmetries. $\Psi^{(+)}_{k_i}$ and $\Psi^{(-)}_{k_f}$ are the initial and final distorted continuum wave functions with the outgoing- (+) and incoming-wave (-) boundary conditions, respectively. U_{se} is the static-exchange potential operator. The wave functions $\Psi^{(+)}_{k_i}$ and $\Psi^{(-)}_{k_f}$ were calculated using the iterative Schwinger variational method [25] in the static-exchange field of the ground state and excited state of the target, respectively. Although the continuum orbitals of the incoming electron are known to be orthogonal to all bound orbitals of the ground-state target, it is not the case for the scattered electron since the continuum wave functions are calculated in the static-exchange field of the open-shell target. For simplicity, these outgoing orbitals were constrained to be orthogonal to all occupied orbitals of the excited target. Further, the transition T matrix as well as the distorted wave functions are single-center expanded. Such expansions are truncated at the cutoff parameters $l_c = 90$ and $m_c = 9$. For the transitions leading to the singlet excited state, the contributions from the higher partial waves were accounted for via the Born-closure procedure [26, 27]. The ground-state configurations of the targets are represented by single-determinant near-HF wave functions. The basis sets used for the HF calculations can be found in [17]. The same basis sets are also used to calculate the wave functions of the lowest unoccupied orbitals (2π in CO, $2\pi_u$ in CO_2 and 4π in OCS) and the vertical excitation energies using an improved virtual orbital approximation [28]. The calculated vertical excitation energies for the singlet and triplet transitions from the O(1s) orbital to the

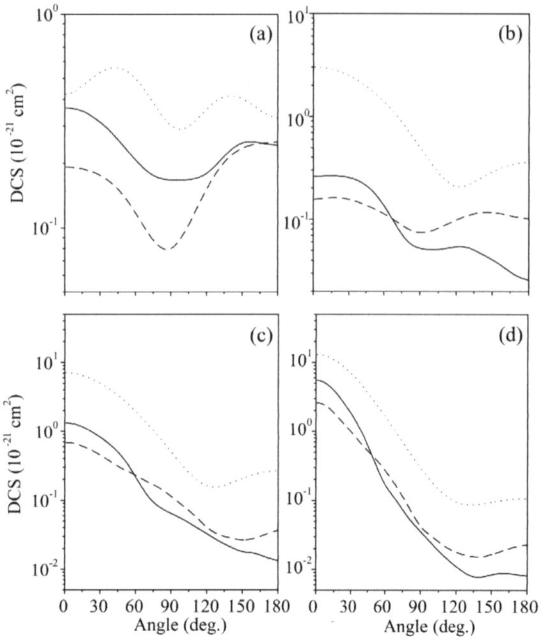

Figure 1. Present DWA DCS for electron-impact $O(1s)^{-1}$ singlet transitions in CO, CO_2, and OCS at (a) 556 eV, (b) 600 eV, (c) 700 eV, and (d) 900 eV. Solid line, results for the $2\sigma \rightarrow 4\pi$ transition in OCS; dashed line, DCS for the $1\sigma_g \rightarrow 2\pi_u$ transition in CO_2; dotted line, results for the $1\sigma \rightarrow 2\pi$ transition in CO.

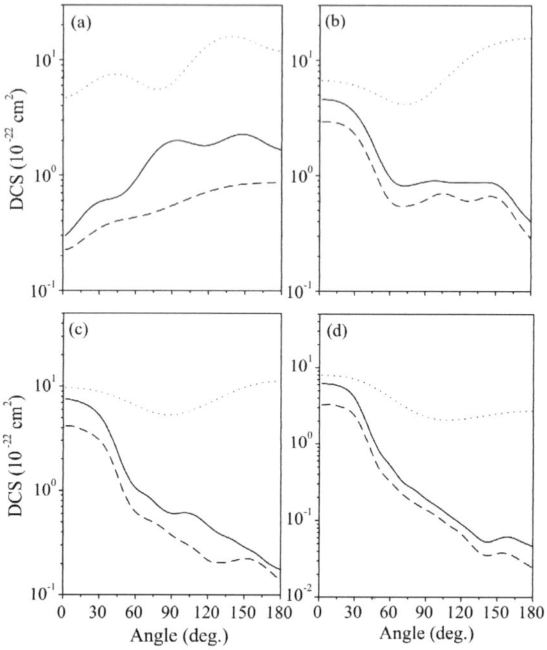

Figure 2. The same as in Fig. 1, but for the triplet transitions.

lowest π orbital, as well as the singlet-triplet energy splitting are shown in Table I, in comparison with some available data in the literature. It is seen that the calculated vertical excitation energies for singlet transitions are about 3% higher than the corresponding experimental values for these molecules, except CO, which is quite reasonable.

3. Results and discussion

Fig. 1 compares the calculated DWA DCS for the singlet transitions: $X\,^1\Sigma^+ \rightarrow^{1,3} \Pi(\,1\sigma \rightarrow 2\pi)$ in CO, $X\,^1\Sigma_g^+ \rightarrow^{1,3} \Pi_u(\,1\sigma_g \rightarrow 2\pi_u)$ in CO_2, and $X\,^1\Sigma^+ \rightarrow^{1,3} \Pi(2\sigma \rightarrow 4\pi)$ in OCS in the (550-900)-eV incident energy range. In Fig. 2, a similar comparison, but for the corresponding triplet transitions, is presented. It is seen in these figures that the DCS calculated for CO are substantially larger than the corresponding results for OCS and CO_2 in the covered energy range. In fact, the observed discrepancies are somehow expected. According to Eq. (3), the transition T matrix depends on both the core and excited valence orbitals. The $1\sigma_g$ orbital of CO_2 is delocalized and thus is expected to be different from the localized O(1s) orbital of OCS and CO. For the latter two targets, despite their O(1s) orbitals are both strongly localized and atomic-like, the lowest unfilled π orbitals of the two targets are certainly very different, since CO is diatomic and OCS triatomic.

For the same reason, the significant differences seen between the DCS calculated for OCS and CO_2 at incident energies of 556 and 600 eV can be justified. Nevertheless at higher incident

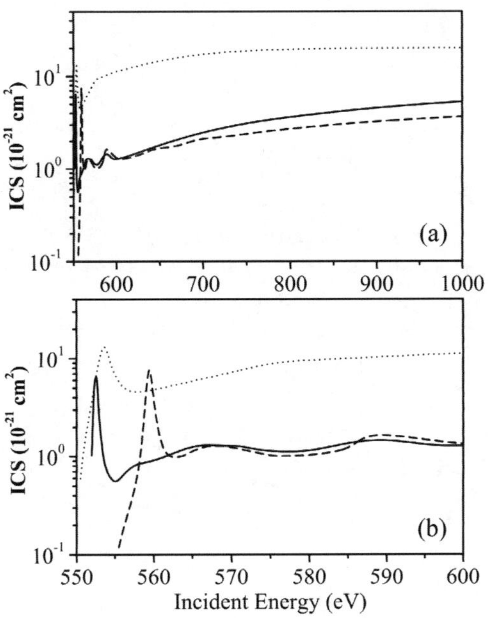

Figure 3. Present DWA ICS for electron-impact $O(1s)^{-1}$ singlet transitions in CO, CO_2, and OCS in the (a) (550-1000)-eV and (b) (550-600)-eV energy ranges. Solid line, results calculated for the $2\sigma \rightarrow 4\pi$ transition in OCS; dashed line, ICS for the $1\sigma_g \rightarrow 2\pi_u$ transition in CO_2; dotted line, results for the $1\sigma \rightarrow 2\pi$ transition in CO.

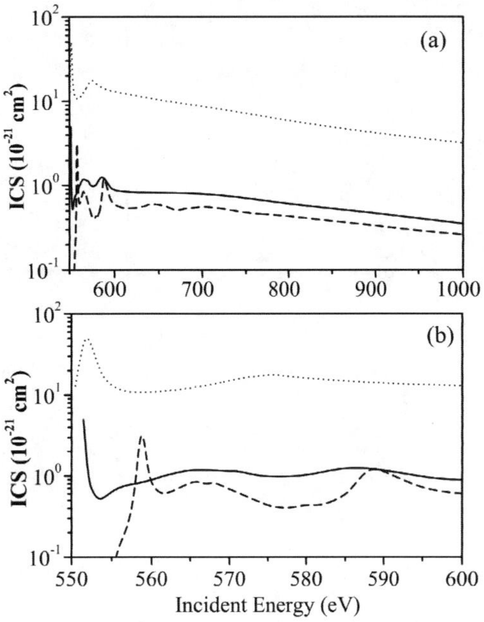

Figure 4. The same as in Fig. 3, but for the triplet transitions.

energies, a fairly good agreement, both qualitative and quantitative, in the DCS calculated for these two targets is seen which is surprising. Maybe, it is because that OCS and CO_2 are both triatomic linear molecules, and are also isoelectronic in the valence shell. Thus, there could be some similarities between the 4π orbital of OCS and the $2\pi_u$ orbital of CO_2 that contribute to the good agreement in calculated DCS for these targets at high incident energies. Therefore, our study suggests that the influence of the localized or delocalized character of the core orbitals in the calculated DCS decreases with increasing energies.

In Figs. 3 we show our DWA ICS calculated in the (550-1000)-eV energy range for electron-impact transitions leading to the singlet excited states: $(1\sigma)^{-1}2\pi$ in CO, the $(2\sigma)^{-1}4\pi$ in OCS, and the $(1\sigma_g)^{-1}2\pi_u$ in CO_2. Calculated ICS for the transitions leading to the corresponding triplet excited states are shown in Fig. 4. As expected, the non-resonant ICS for the excitation of one O(1s) electron in CO are substantially larger than those for OCS and CO_2. In addition, for both singlet and triplet transitions, fairly good agreement is seen between the non-resonant ICS calculated for OCS and CO_2. Near the excitation thresholds, resonance-like structures are seen in the DWA ICS for the three targets. For singlet transitions, a strong peak located at around 552 eV and two weak peaks around 567 and 588 eV, respectively, are seen in the results of OCS. Similar resonance structures are also seen in the calculated ICS of CO_2, except the position of the the strongest peak is located at about 559 eV. On the other hand, only two resonances are seen in the ICS for CO: one sharp feature located at about 553 eV and one shoulder at around 577 eV. For triplet transitions, the resonance structures of the three molecules are quite similar

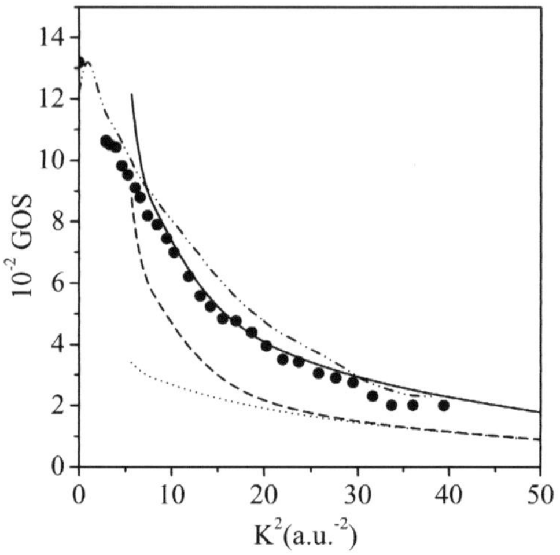

Figure 5. GOS for electron-impact core-excitation of CO_2 at 1300 eV. Dotted line, present calculated results for the $1\sigma_u \rightarrow 4\pi_u$ singlet transition; dashed line, DWA GOS for the $1\sigma_g \rightarrow 2\pi_u$ singlet transition; Solid line, present summed GOS; dashed-dotted-dotted line, GOS of Eustatiu *et al.* [12] using the FBA; full circles, experimental results of GOS of of Eustatiu *et al.* [12].

to their counterparts in the singlet transitions, except their positions are slightly shifted.

The resemblance between the resonance structures in OCS and CO_2 seems to reinforce the idea that the low-lying unfilled orbitals of the two valence-shell isoelectronic systems are quite similar with each other. In contrast, it is expected that the low-lying excited orbitals of CO are significantly dissimilar to those of OCS and CO_2. Also, eigenphase analyses (not shown) have revealed that the strongest resonance feature for the singlet transition in CO_2 is due to the occurrence of a $k\pi_u$ shape resonance. Similar peaks seen in the the ICS of CO and OCS are due to a shape resonance in the $k\pi$ scattering channel. For other structures, no eigenphase analyses were performed.

In Fig. 5, we compare our GOS calculated at 1300 eV for the O(1s) core-excitation in CO_2 with the only existing theoretical and experimental data of Eustatiu *et al.* [12]. Due to the small energy difference between the $1\sigma_g \rightarrow 2\pi_u$ and $1\sigma_u \rightarrow 2\pi_u$ singlet transitions, the experimental GOS are in fact the sum of the contributions of the two transitions. Therefore, specifically for this energy, we have also calculated GOS for the $1\sigma_u \rightarrow 2\pi_u$ singlet transition. The calculated results of GOS for the two transitions are also presented separately in Fig.5. It is seen that the GOS for the $1\sigma_u \rightarrow 2\pi_u$ singlet transitions grow much faster than those for the $1\sigma_u \rightarrow 2\pi_u$ transition towards the small momentum transfer (K) region. This is because the former transition is dipole-allowed and the latter is dipole forbidden. Besides, at large momentum transfers, the GOS of both transitions are very similar. The sum of them agree well with the experimental and theoretical data of Eustatiu *et al.* [12], which is encouraging.

In summary, the present work reports an application of the DWA to study core-level excitations from the O(1s) orbital to the lowest lying unoccupied valence orbitals in CO, OCS, and CO_2. The comparison of our DWA GOS for CO_2 with the existing theoretical and experimental data shows good agreement, which confirm the reliability of our method. Moreover, the similarity between the calculated DCS and ICS for OCS and CO_2 at high incident energies as well as the resemblance between the resonance structures in the ICS of these two systems seems to indicate the similarity of the lowest-lying unoccupied orbitals (4π in OCS and $2\pi_u$ in CO_2). On the other hand, the discrepancy seen in the low incident energies can partially be attributed to the atomic-like character of the 2π orbital in OCS and molecular character of the $1\sigma_g$ orbital of CO_2.

4. Acknowledgments

This research was partially supported by CNPq, FAPESP, FUNPAR and Fundação Araucária. Four of us (FA, KTM, OAP and HLO) thank CNPq for scholarships.

5. References

[1] A.P. Hitchcock, J. Electron Spectrosc. Rel. Phenom. **112** 9 (2000).
[2] K. Siegbahm, C. Nordling, G. Johansson, J. Hedman, P.F. Heden, K. Hamrin, U. Gelins, T. Bergmark, L.O.Werme, R. Manne, Y. Baer, ESCA Applied to Free Molecules, North-Holland, Amsterdan, (1969).
[3] J. Stöhr, in: R. Prins, (Ed.), X-Ray Absorption, Principal Application, Techniques of EXAFS, SEXAFS and XANES, Wiley, New York, (1986).
[4] J. J. Neville, T. Tyliszczak, and A. P. Hitchcock, J. Electron Spectrosc. Rel. Phenom. **101** 119 (1999).
[5] S. Masuda, T. Gejo, M. Hiyama, and N. Kosugi, J. Electron Spectrosc. Rel. Phenom. **144** 215 (2005).
[6] Y. Hikosaka, T. Kaneyasu, E. Shigemasa, Y. Tamenori, and N. Kosugi, Phys. Rev. A **75** 042708 (2007).
[7] T. Gejo, M. Machida, K. Honma, E. Shigemasa, E. Nakamura, N. Kosugi, and Y. Tamenori, J. Electron Spectrosc. Rel. Phenom. **156** 274 (2007).
[8] I. L. Bradeanu, R. Flesch, N. Kosugi, A. A. Pavlychev, and R. Rühl, Phys. Chem. Chem. Phys. **8** 1906 (2006).
[9] T. X. Carroll and T. D. Thomas, J. Chem. Phys. **92** 7171 (1990).
[10] T. X. Carroll, De Ji, and T. D. Thomas, J. Electron Spectrosc. Rel. Phenom. **51** 471 (1990).
[11] M.P. Miranda, C.E. Bielschowsky, and M.A.C. Nascimento, J. Phys. B **28** L15 (1995).
[12] I.G. Eustatiu, T. Tyliszczak, A.P. Hitchcock, C.C. Turci, A.B. Rocha, and C.E. Bielschowsky, Phys. Rev. A **61** 042505-1 (2000).
[13] J. Wang, G. Cooper, D. Tulumello, and A.P. Hitchcock, J. Phys. Chem. **109** 10886 (2005).
[14] V. Brems, B. M. Nestmann, and S. D. Peyerimhoff, Chem. Phys. Lett. **287** 255 (1998).
[15] M. Hiyama and N. Kosugi, J. Electron Spectrosc. Rel. Phenom. **144** 1223 (2005).
[16] M. Hiyama and N. Kosugi, J. Phys. B **39** 1797 (2006).
[17] T. Kroin, S.E. Michelin, A.S. Falck, F. Arretche, I. Iga and M.-T. Lee, Phys. Rev. A **68** 012701 (2003).
[18] S.E. Michelin, A. de Campos, L.S.S. da Silva, A.S. Falck, E.A.Y. Castro, O. Pessoa, H.L. Oliveira, M.-T. Lee, Chem. Phys. **293**
[19] S.E. Michelin, O. Pessoa, H.L. Oliveira, E. Veiteinheimer, A.M.S. Santos, M.M. Fujimoto, I. Iga I, M.-T. Lee, Phys. Rev. A **72** 022730 (2005).
[20] A.W. Fliflet and V. McKoy, Phys. Rev. A **21** 1863 (1980).
[21] L.M. Tao and V. McKoy, J. Phys. B **15** 3971 (1982).
[22] M.-T. Lee, L.M. Brescansin and M.A.P. Lima, J. Phys. B **23** 3859 (1990).
[23] M.-T. Lee, S.E. Michelin, T. Kroin, L.E. Machado and L.M. Brescansin, J. Phys. B **28** 1859 (1995).
[24] A.R. Edmonds, *Angular Momentum and Quantum Mechanics*, (Princenton University Press, Princenton, NJ, 1960).
[25] R.R. Lucchese, G. Raseev, V. McKoy, Phys. Rev. A **25** 2572 (1982).
[26] L.E. Machado, L. MuTao, L.M. Brescansin, M.A.P. Lima, and V. McKoy, J. Phys. B **28** 467 (1995).
[27] L.E. Machado, E.P. Leal, L. MuTao, and L.M. Brescansin, THEOCHEN:J. Mol. Struct. **335** 37 (1995).
[28] W.A. Goddard and W.J. Hunt, Chem. Phys. Lett. **24** 464 (1974).

Low Energy (0-12 eV) Electron Interaction with Gas Phase Building Blocks of DNA/RNA

Ilko Bald[1,a], Janina Kopyra[1,3], Constanze König[1], Helga D. Flosadottir[2], Oddur Ingolfsson[2], Eugen Illenberger[1]

[1]Physikalische und Theoretische Chemie, Institut für Chemie und Biochemie, Freie Universität Berlin, Takustrasse 3, D-14195 Berlin, Germany

[2]Department of Chemistry, Faculty of Science, University of Iceland, Science Institue, Dunhaga 3, 107 Reykjavik, Iceland

[a]ibald@zedat.fu-berlin.de

Abstract. We review recent results on dissociative electron attachment (DEA) to gas phase D-ribose, tetraacetyl-D-ribose (TAR) and dibutylphosphate (DBP), which serve as model compounds for the DNA or RNA backbone. New results are presented on negative ion formation in D-ribose probed by matrix-assisted laser desorption/ionisation time-of-flight (MALDI-TOF) mass spectrometry. The two methods reveal that the transient D-ribose anion $R^{-\#}$ decomposes in the same way like the deprotonated D-ribose molecule $[R-H]^-$, i.e. by abstraction of different numbers of water and formaldehyde units. In DEA the TNI $R^{-\#}$ is generated at very low energies close to 0 eV most likely through a vibrational feshbach resonance. The fragmentation pattern and the characteristic resonances of D-ribose are preserved in TAR, where a furanose is bound to four acetyl groups. The presence of an acetyl group leads additionally to fragmentation through a shape resonance. Shape resonances were also observed in DBP, followed by C-O and P-O bond breaking.

1. Introduction

We present recent results on the interaction of low energy electrons (LEEs) with the components of the DNA backbone in an effort to unravel the molecular mechanism how low energy electrons (LEEs) damage DNA. We consider biomolecular systems that may serve as appropriate models to probe the response of the sugar and the phosphate moiety in DNA towards the attack of LEEs. It appears that in order to assess the role of a particular DNA component it is not sufficient to study the isolated molecule. Instead one has to find appropriate molecular systems that mimic its coupling within DNA.

The interaction of high energy radiation (α, β, γ) abundantly generates secondary electrons along the ionisation track [1]. In the meantime it is well recognised that reactions induced by these LEEs with the vital components of a cell (DNA, proteins, water) is an important contributor in radiation damage of living cells. Within this context, reactions *directly* induced in DNA are referred to as *direct damage*. With regard to the molecular mechanisms behind direct radiation damage a great number of experimental and theoretical works has been performed within the last few years [2], and the state of the art can briefly be summarised as follows:

[3] permanent address: Department of Chemistry, University of Podlasie, 08-110 Siedlce, Poland

Exposure of plasmid DNA to free electrons at energies already *below* the ionisation threshold leads to single strand breaks (SSBs) and double strand breaks (DSBs) [3]. These findings contrast the traditional notion which considered only electrons *above* the ionisation threshold to be involved in strand breaks. The number of strand breaks per incident electron flux shows a resonant behaviour with the electron energy. Both SSBs and DSBs are observed within a resonant feature in the energy range between 6-15 eV [3] while electron irradiation within a low energy resonance (in the range 0-4 eV) only induces SSBs [4]. These findings suggest that resonant capture of electrons at particular DNA sites may be the initial step towards SSBs and DSBs.

The interaction of LEEs with the isolated gas phase DNA bases thymine (T), cytosine (Cy), adenine (A) and guanine (G) studied in crossed electron- molecular beam experiments [5-10] showed that all the DNA bases undergo dissociative electron attachment (DEA) via two resonant features located between 0-4 eV and 6-12 eV, similar to the energy dependence of strand breaks in plasmid DNA. While processes within the high energy feature lead to the loss of H^- and larger fragment ions [6], partly associated with the deterioration of the cyclic structure, DEA via the low energy feature exclusively abstracts a neutral hydrogen atom from the N sites thereby generating the closed shell DNA base anion [7,8]. This hydrogen loss is subjected to a remarkable site selectivity as (in T) electrons of 1 eV energy exclusively induce loss of H from the N1 site while at 1.8 eV this occurs preferentially from the N3 site [8].

Experiments on self assembled monolayers of short oligonucleotides [11] and oligonucleotides immobilised on microarrays [12] revealed that the DNA bases act as antennas for electrons at subexcitation energies thereby also inducing damage of the oligonucleotide [12].

Based on these findings an electron transfer mechanism for DNA strand breaks was proposed by theoretical studies [13,14] modelling a section of DNA composed of cytosine, sugar and the phosphate group. These calculations predict a low-lying anionic potential energy surface which connects the initial π^* anion state of the base to a σ^* state in the backbone. An electron captured by a DNA base may thereby be transferred to the backbone leading to rupture of the C-O bond between the phosphate and the sugar.

On the other hand, gas phase studies on thymidine (a thymine coupled to a deoxyribose molecule by the glycosidic N1-C1 bond) [15] did not show any electron transfer from thymine to the sugar moiety via the low energy resonance, at least no electron transfer associated with DEA. The low energy resonance only resulted in the H loss from the N3 position. These findings do not necessarily contradict the above electron transfer mechanism as the presence of the phosphate group may modify the electron transfer behaviour of the system. It demonstrates, however, that coupling of the molecule in the DNA network significantly modifies its properties with respect to DEA. While in T the loss of neutral hydrogen from the N1 position is the dominant DEA reaction in the free molecule, DEA is completely suppressed once the deoxyribose is coupled at N1 [15]. This observation is corroborated by methylation experiments indicating that the loss of H is completely blocked, once the corresponding hydrogen atoms are replaced by the methyl group [8]. These results clearly demonstrate that in order to study the role of individual DNA components it is not sufficient to probe the isolate molecule but one has to find appropriate molecular systems that mimic its coupling in DNA.

An alternative mechanism for DNA strand breaks is based on proton transfer triggered by an excess electron [16]. DFT calculations on the nucleotide of cytosine predict that Cy hydrogenated at the C6 position (Cy+H) can strongly bind an excess electron to form the closed shell moiety $(Cy+H)^-$ with the excess charge localized at the C-6 position of $(Cy+H)^-$. A proton transfer process then induces a barrier free sugar-phosphate C-O bond cleavage. It should be noted that these calculations refer to the electronic ground state of the system, while the electron transfer model from above [13] treats the system initially as a resonant state, i.e. the DNA base captures a free ballistic electron to form a transient negative ion state which is embedded in the detachment continuum.

On the background of the present experimental material and the different theoretical models the evolution of an excess charge initially localised on a DNA base is hence not resolved. The question then arises on the sensitivity of the sugar and the phosphate unit towards low energy electrons. A DFT treatment of the thymine nucleotide [17] in fact predicts low lying resonant states localised directly on the phosphate moiety. An exploration of the adiabatic and vertical (C-O) potential energy surfaces (including solvation by water molecules and the presence of a Na^+ counter ion at the phosphate) supports the picture that LEEs transiently captured at the phosphate can induce SSBs via rupture of the sugar-phosphate (C-O) bond.

Here we review recent results on DEA to ribose [18], tetraacetyl-D-ribose (TAR) [19] and dibutylphosphate (DPB) [20]. The latter two systems can be considered as reasonable surrogates to probe the behaviour of the deoxyribose and phosphate moiety in DNA towards the interaction of low energy electrons. In combination with new experiments on negative ions generated through matrix assisted laser desorption and ionisation (MALDI) a more comprehensive picture of negative ion formation and decomposition in D-ribose is obtained.

2. Experiments

Dissociative electron attachment (DEA) is studied using a crossed electron molecular beams arrangement [21], which is housed in a high-vacuum chamber. The electron beam is generated with a trochoidal electron monochromator (TEM) [22] allowing an electron energy resolution of ≈ 100 meV at a current of 5-20 nA. The molecules of interest are evaporated from a small vessel inside the vacuum system by heating the whole chamber with two halogen bulbs to moderate temperatures not higher than 370 K. In the reaction volume the evaporative molecular beam crosses the electron beam, and the generated anions are extracted by ion optics and mass analysed and detected with a quadrupole mass spectrometer. The energy scale is calibrated using the formation of metastable SF_6^- from SF_6 at 0 eV. The flow of SF_6 gas was switched off prior to each measurement.

MALDI-TOF experiments have been performed in the Reykjavik laboratory with a REFLEX IV (Bruker Daltonics) instrument that is equipped with a time-of-flight (TOF) Reflectron type mass spectrometer and a pulsed N_2 laser (337 nm, 10 Hz, 400 μJ/pulse) [23]. Ion extraction is pulsed with 400 ns delay time. For each spectrum the samples were irradiated at different spots to compensate for inhomogeneities of the sample. In-source decay (ISD) spectra were measured in reflectron mode using 300 laser shots. For post-source decay (PSD) the deprotonated D-ribose was selected at a gate and its fragmentation was measured in six mass segments. For each segment the reflection voltage is stepped down to ensure for collection and detection of all fragment masses, and 500 laser shots per segment were collected. Samples were prepared by deposition of 0.5 μL of a 3.5 mM solution of matrix (Bis-benzimide ($C_{25}H_{24}N_6O$)) in methanol on a polished stainless steel sample carrier and allowed to dry in air. Afterwards 0.5 μL of 0.13 M D-ribose solution in methanol was spotted on the matrix and allowed to dry.

5-^{13}C-D-ribose and *C*,1-D-D-ribose were obtained from Cambridge Isotopes Laboratories, Inc., and all other samples were purchased from Sigma Aldrich. All substances were used as delivered.

3. Results and Discussion

All investigated compounds serve as models for the DNA or RNA backbone. The isolated monosaccharide D-ribose ($C_5H_{10}O_5$) is a rough model for the sugar unit in RNA (and DNA), tetraacetyl-D-ribose is used as an improved model for a furanose sugar coupled within a more complex molecular environment, and dibutylphosphate mimics a neutral phosphate group bound to two aliphatic carbon chains. Our findings indicate that all compounds are sensitive towards low energy electron attachment and show characteristic ion yield curves. The anion formation and fragmentation is ascribed to the dissociative electron attachment (DEA) mechanism [21], *viz.*:

$$ABC + e^- \rightarrow ABC^{-\#} \rightarrow AB + C^- \qquad (1)$$

An electron of defined energy is resonantly captured by the neutral molecule to form a transient negative ion (TNI). The TNI $ABC^{-\#}$ is basically unstable towards autodetachment of the extra electron, however, if the lifetime exceeds one vibrational period it may dissociate into a stable anion and neutral fragment. The resonant character of the process is determined in the ion yield curve of a given fragment ion, for instance C^-. The production of a fragment ion may proceed either directly within the lifetime of the resonance, i.e. usually 10^{-15} – 10^{-12} s, or indirectly via the formation of metastable intermediates. The DEA experiment does not allow distinguishing these processes whereas some details of the fragmentation dynamics are revealed by MALDI experiments.

DFT calculations (section 3.1.1.) on stable anionic products generated through DEA to D-ribose and MALDI mass spectra of D-ribose (section 3.1.2.) reveal slightly different reaction pathways and structures of product anions than previously suggested in [18].

3.1. D-ribose

3.1.1. DEA spectra. D-ribose captures electrons already at very low energies close to 0 eV and decomposes subsequently into various fragment anions [18]. The most effective dissociation channel is loss of one or two water molecules leading to formation of $C_5H_8O_4^-$ (132 amu) and $C_5H_6O_3^-$ (114 amu), respectively [18]. But also the carbon ring is decomposed resulting mainly in generation of $C_4H_5O_3^-$ (101 amu), $C_3H_4O_2^-$ (72 amu) and $C_2H_3O_2^-$ (59 amu).

Unambiguous assignment of decomposition products is achieved using the isotope labelled analogues 5-^{13}C-D-ribose, 1-^{13}C-D-ribose and *C*,1-D-D-ribose (figure 1). For instance, the 101 amu fragment can also be assigned to $C_5H_9O_2^-$ that contains all five carbon atoms. However, the DEA spectra in figure 1 show that in 5-^{13}C-D-ribose the signal remains at 101 amu, whereas it vanishes completely in 1-^{13}C-D-ribose and *C*,1-D-D-ribose, respectively, and shifts to 102 amu (left column of figure 1). Moreover, it follows that production of $C_4H_5O_3^-$ proceeds via selective excision of C-5. The experiment does not deliver information on the neutral decomposition products, but the most likely composition is $CH_2O + H_2O + H$, since all products can be formed with low bond-breaking barriers from the sugar. However, the thermodynamics are more favorable for other channels (sum of heats of formation of neutral products is $\Sigma_{neutr.prod.}\Delta_fH^0 = -140$ kJmol^{-1} [24]) like $CO_2 + 2\ H_2 + H$ ($\Sigma_{neutr.prod.}\Delta_fH^0 = -176$ kJmol^{-1} [24,18]). Nevertheless the latter reaction is unlikely due to large activation barriers. Consequently the carbon atom of the generated formaldehyde molecule, CH_2O, is C-5 (figure 2).

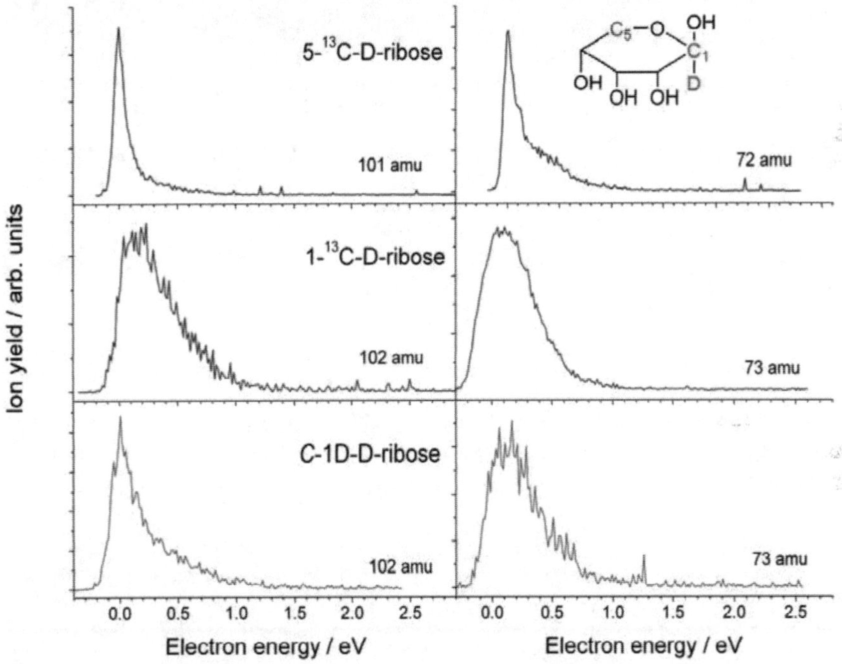

Figure 1. DEA spectra of selected fragments of different isotope labelled D-ribose molecules: The left column shows the 101 amu fragment anion, which is shifted to 102 amu in 1-^{13}C-D-ribose and *C*,1-D-D-ribose. The right column displays the 72 amu fragment anion, which is shifted to 73 amu in 1-^{13}C-D-ribose and *C*,1-D-D-ribose. Adapted from [18].

Basically the molecular structure of the 101 amu fragment may be a 5-ring structure containing one oxygen atom. However, DFT calculations on different stable anionic structures for $C_4H_5O_3^-$ reveal that only an open chain anion, that is stabilized by a hydrogen bond (as is shown in figure 2) can be formed at threshold [25].

The fragment anion at 72 amu (shown in the right panel of figure 1) is ascribed to the following reaction: e^- ($\approx$ 0 eV) + $C_5H_{10}O_5 \rightarrow C_5H_{10}O_5^{-\#} \rightarrow C_3H_4O_2^-$ (72 amu) + 2 CH_2O + H_2O. The site selectivity is similar to the 101 amu fragment since the m/Z 72 signal is almost completely shifted in 1-^{13}C-D-ribose and *C*,1-D-D-ribose to 73 amu but almost no signal is observed at 73 amu in 5-^{13}C-D-ribose. Consequently the only difference to the 101 amu fragment anion is presumably the generation of an additional formaldehyde molecule that is most likely formed from C-4 (figure 2). However, from thermodynamics the production of neutral CO_2, methanol (CH_3OH) and H_2 is more favourable ($\Sigma_{neutr.prod.}\Delta_f H^0$ = -595 kJmol^{-1} [24]) compared to $\Sigma_{neutr.prod.}\Delta_f H^0$ = -518 kJmol^{-1} for production of formaldehyde and water.

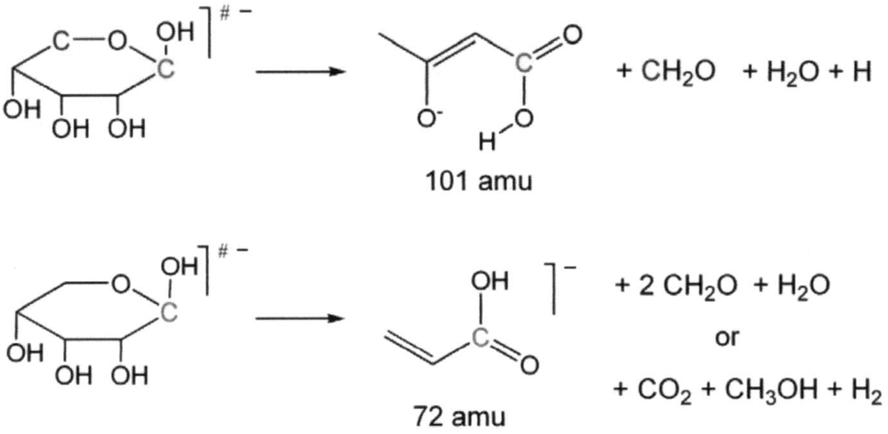

Figure 2. Formation of the 101 amu and 72 amu fragment ions from the D-ribose transient anion. According to recent DFT calculations [25] the reaction pathways shown here are exothermic; they differ slightly from the reaction products originally proposed in [18].

3.1.2. MALDI-TOF spectra. Figure 3 displays MALDI-TOF mass spectra of anions produced from D-ribose. In ISD (upper panel of figure 3) the main fragmentation reactions are:

$C_5H_{10}O_5 \rightarrow$ (R-H)$^-$ (149 amu) + H (2)
$\rightarrow C_5H_5O_4^-$ (129 amu) + H_2O + H_2 + H (3)
$\rightarrow C_4H_4O_3^-$ (100 amu) + H_2O + CH_2O + H_2 (4)
$\rightarrow C_3H_3O_2^-$ (71 amu) + 2 CH_2O + H_2O + H (5)
$\rightarrow C_2H_2O_2^-$ (58 amu) + H_2O + $C_3H_5O_2$ + H (6)

The ISD spectrum is due to ionization and fragmentation inside the ion source, i.e. prompt decay in the plume up to early metastable decay within the 400 ns before ion extraction. Unlike the DEA experiments described above the conditions in the MALDI plume are far from single collision conditions [26], that is multiple collisions are likely and the observed product ions have a chance to relax through collisions. The ionization mechanism in MALDI is not fully understood to date, but deprotonation of the analyte molecules leading to a closed shell anion is considered to be the most important process. Additionally electron capture processes may be involved [26,27]. The presence of low energy photoelectrons was shown previously through the production of SF_6^- from SF_6 [27], which was introduced into the vacuum system of a MALDI source combined with a FT-ICR mass spectrometer. A similarity of the fragmentation patterns in MALDI ISD and DEA processes is obvious, i.e. release of neutral water and formaldehyde molecules from the parent anion.

In most cases the signals in MALDI ISD are shifted by one mass unit (100, 71, 58 amu) compared to DEA spectra (mainly 101, 71, 72 and 59 amu). One exception is the signal at 129 amu, which is accompanied by small contributions of 130 and 131 amu. The deprotonated parent ion $[R\text{-}H]^-$ at 149 amu was not observed at all in DEA.

The DEA experiments are controlled by single-collision conditions, i.e. the reactive transient negative ion is an open shell radical anion that decays into one (neutral or anionic) radical species. In other words the whole system always contains an excess electron and relaxes into a thermodynamically stable system with respect to the TNI. In contrast to that the deprotonated D-ribose is a closed shell anion and the fragmentation is controlled by the *neutral* products leading finally to closed shell fragment anions having one hydrogen less than the ions observed in DEA. If we additionally consider electron capture processes that may take place in ISD, it has to be noted that the in-plume conditions in MALDI allow generation of thermodynamically more stable products, i.e. closed shell anions. For instance abstraction of two formaldehyde molecules and one water molecule from the TNI $R^{-\#}$ generates the radical anion $C_3H_4O_2^-$ (72 amu) that presumably possesses two conjugated double bonds with the extra electron residing in a π^* orbital. The radical anion may still be reactive if it is allowed to collide, and may thus loose a hydrogen atom in the course of collisions with surrounding molecules - for instance matrix molecules - resulting in the anion $C_3H_3O_2^-$ with mass 71 amu as it is observed in ISD and PSD.

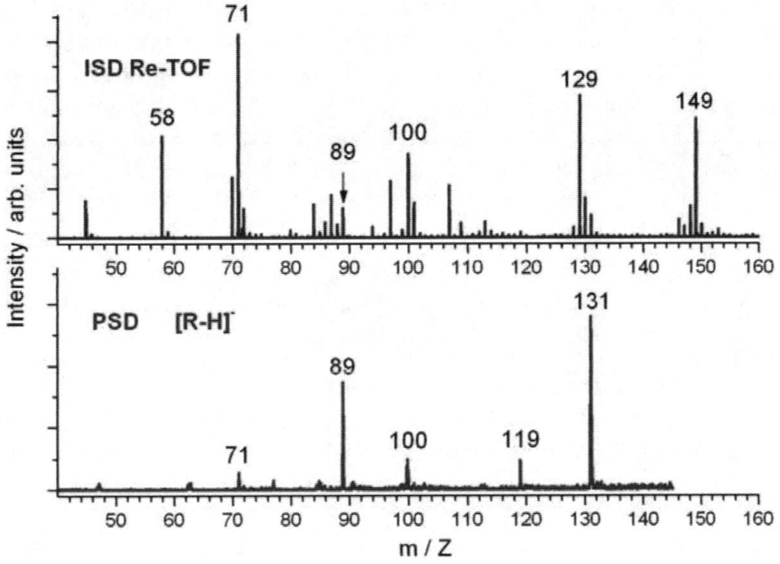

Figure 3. MALDI-TOF in-source decay (ISD) spectrum (upper panel) and post-source decay (PSD) spectrum (lower panel) of D-ribose.

The same considerations are valid for the PSD spectrum of the closed shell deprotonated D-ribose molecular anion $[R\text{-}H]^-$ (lower panel of figure 3). $[R\text{-}H]^-$ decays into the following products:

$$C_5H_9O_5^- \text{ (149 amu)} \rightarrow C_5H_7O_4^- \text{ (131 amu)} + H_2O \quad (7)$$
$$\rightarrow C_4H_7O_4^- \text{ (119 amu)} + CH_2O \quad (8)$$
$$\rightarrow C_4H_4O_3^- \text{ (100 amu)} + H_2O + CH_2O + H \quad (9)$$
$$\rightarrow C_3H_5O_3^- \text{ (89 amu)} + 2\,CH_2O \quad (10)$$
$$\rightarrow C_3H_3O_2^- \text{ (71 amu)} + 2\,CH_2O + H_2O \quad (11)$$

All these anions occur in DEA as well, but with one hydrogen atom more and in different intensity ratios. Thus the *neutral* decomposition products are identical in DEA and MALDI-PSD.

The comparison with free electron attachment shows that the TNI of D-ribose $R^{-\#}$ as well as the deprotonated D-ribose $[R\text{-}H]^-$ decompose in a characteristic manner, i.e. by abstraction of different numbers of neutral water and formaldehyde molecules. The abstraction of one and two formaldehyde molecules (in addition to one H_2O molecule; resulting in 101/100 amu and 72/71 amu, respectively) proceeds site selectively from C-5 and C-4 (figure 2).

3.1.3. Mechanism of ion generation in DEA. The DEA reactions discussed above occur with electrons having essentially no energy. The usual mechanism of electron attachment at low energies is described via the formation of shape resonances, i.e. the electron occupies a virtual orbital of the neutral molecule without changing the initial electronic configuration. However, the energy of σ* orbitals of the saturated sugar molecule is too high to be accessible by electrons with energies close to zero eV. In fact recent calculations on electron scattering on D-ribose [28] showed that no shape resonances occur at low energies, but higher energy signals in the ion yield curves of the 72 amu and 59 amu fragments (not shown here) may be explained by the presence of shape resonances [28].

A possible explanation of the threshold signals is the initial formation of a vibrational feshbach resonance [29], which is supported by the large dipole moment of the sugar molecules. For such a doorway mechanism certain conditions need to be fulfilled [30,31]: (1) A dipole bound state must exist, where the excess electron occupies a very diffuse Rydberg-like orbital, (2) a valence anionic state with a positive electron affinity must be present, (3) the coupling of the dipole bound and the valence anionic states must be efficient, and (4) the valence state has to dissociate into more stable products. A recent *ab initio* study on D-fructose [30] indeed showed that a dipole bound state having a binding energy of at least 5 meV as well as a stable valence state exist. The latter is an open-chain distonic radical anion, which is predicted to be the precursor for dissociation in D-fructose. The question if the coupling between these two states is strong was not fully answered in [30] due to the complexity of the sugar molecule, but it is expected that the presence of an additional electron lowers the bond-breaking barriers enough to enable electron transfer from the dipole bound state to the open-chain valence state.

3.2. Tetraacetyl-D-ribose (TAR)

The rich fragmentation of the transient D-ribose anion is remarkable; however, the structure of the sugar unit in DNA and RNA is considerably different from the isolated gas phase sugar. The question arises, which resonances and fragmentation reactions are preserved in a furanose sugar that is also bound to a more complex molecular network, in other words which does not possess free hydroxyl groups. Hence we investigated DEA of the furanose form of tetraacetyl-D-ribose [19] (TAR, figure 4a-c).

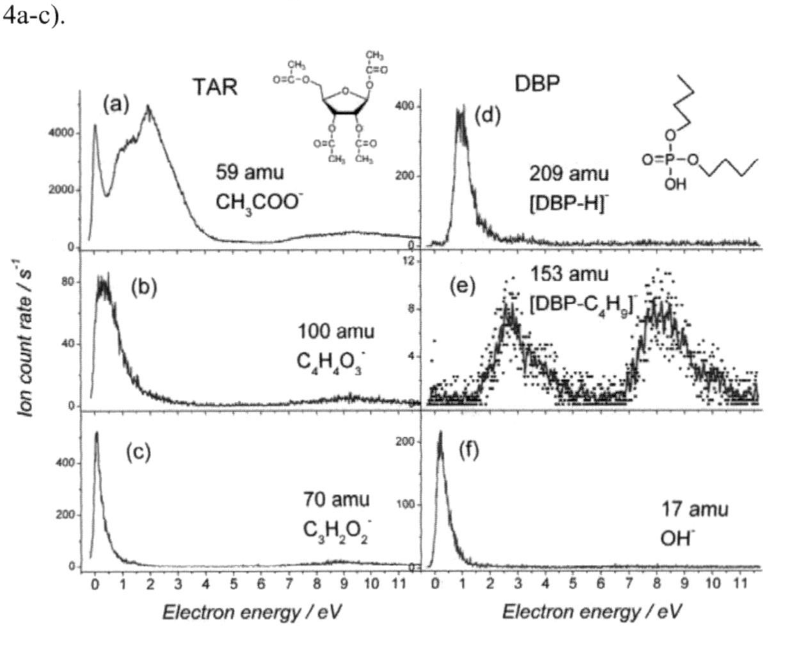

Figure 4. DEA spectra of selected fragment anions of tetraacetyl-D-ribose (TAR, left column) and dibutylphosphate (DBP, right column). Adapted from [19] and [20].

The most abundant fragment anion is the acetate anion at 59 amu. It is generated within three different energy regions: (1) Within a sharp peak close to zero eV, (2) between 1 and 3 eV from at least two overlapping resonances, (3) with lower intensity between 7 and 11 eV. A comparison with the free D-ribose molecule shows that only the 1-3 eV signal is a new feature in the spectra of TAR. It is due to π* shape resonances located at the exocyclic carboxyl groups [19], which are not present in the free sugar. Indeed a series of lower mass fragments (100 amu, 70 amu and 84 amu) that correspond to the sugar ring are generated only at threshold and with lower cross sections between 7 and 11 eV. The π* shape resonance is also observed through generation of higher mass fragment ions (215 amu, 161 amu, 154 amu, 119 amu and 113 amu, not shown here, see [19]), which are produced exclusively at 1–3 eV.

These observations confirm that the threshold signals obtained in the free sugars are preserved also in a five-member ring sugar that is bound to other functional groups. A frequently used surrogate for the sugar in DNA is tetrahydrofuran (THF) [32]. DEA studies on gas phase THF showed that the pure carbon ring has much lower electron attachment cross sections and does not lead to fragmentation at energies below 7 eV [32]. Therefore the presence of several polar groups (hydroxyl or acetate) leads to threshold signals, most likely due to the higher dipole moment resulting in dipole supported states. However the doorway mechanism discussed above for the free D-ribose molecules requires an open chain parent ion as transition state [30]. In TAR C-1 is blocked with an acetate group, i.e. the formation of a chain-form is not possible. Thus the dissociation has to proceed via a cyclic valence state.

In comparison with free D-ribose the π* shape resonances at low energies in TAR from the exocyclic groups appear *in addition* to the threshold signals without suppressing them. This is also expected if a phosphate group is present (see next section), where π* shape resonances occur as well. The high energy signals between 7 and 11 eV correspond to the high energy shape resonance described above for D-ribose [28].

3.3. Dibutylphosphate (DBP)

To model the phosphate group that is bound to two sugar units in DNA and RNA we used dibutylphosphate [20] (figure 4d-f). The parent ion after loss of hydrogen is observed from different resonances between 0.5 eV and 4 eV with a maximum at 1 eV. These signals are assigned to π* shape resonances with temporary occupation of a π* orbital of the phosphate group.

An abstraction of the whole butyl group due to a C-O bond cleavage resulting in a fragment ion with mass 153 amu is observed from two broad resonances at 2-4 eV and 7-10 eV. This reaction corresponds directly to a strand break in DNA. The low energy feature is also assigned to a π* shape resonance located at the phosphate group, and the higher energy resonance is most likely due to a core excited resonance at the carbon chain.

$[DBP\text{-}H]^-$ and $[DBP\text{-}C_4H_9]^-$ arise from a single bond cleavage. Additionally OH^- is easily generated by rupture of a single P-O bond (figure 4f). Multiple bond breaking was observed leading to fragment anions PO^-, PO_3^- and $H_2PO_3^-$ (not shown here, see [20] for details).

3.4. Conclusions

The results on DEA of different model compounds for the DNA or RNA backbone indicate that the backbone itself is able to capture electrons leading to manifold fragmentation reactions with considerable cross sections. The fragmentation of D-ribose negative ions is mainly governed by the neutral decomposition channels as was shown in a comparative study of DEA and MALDI. Fragment anions observed in DEA correspond to the fragments detected in MALDI ISD indicating that fast fragmentation prevails in DEA rather than metastable decay processes. Consequently it is suggested that strand break formation by low energy electrons is also possible by direct electron attachment to the phosphate group or the sugar unit, especially at energies below 4 eV, where only single strand breaks have been observed [4]. To arrive at a global model that fully describes the interaction of LEEs with DNA the next step will be the experimental investigation of larger subunits containing the nucleobases, sugar and phosphate units. Gas phase studies are particularly valuable to reveal the intrinsic properties of these building blocks. This knowledge is vitally important for a proper interpretation of results from more complex systems in condensed phase.

Acknowledgement
Work supported by the Deutsche Forschungsgemeinschaft (DFG) and by the Freie Universität Berlin, by the Icelandic Centre for Research (RANNIS) and by the University of Iceland Research Fund. IB is

a fellow of the Studienstiftung des Deutschen Volkes and acknowledges support for a visit to Reykjavik by the COST Action P9; Radiation Damage in Biomolecular Systems (RADAM). JK acknowledges support from polish scientific funds for the years 2005-2007 by the grant No. 3T09A11129 and support for a visit to Berlin by the EU program COST P9.

References

[1] Cobut V, Fongillo Y, Patau J P, Goulet T, Fraser M-J and Jay-Gerin J-P 1998 *Radiat. Phys. Chem.* **51** 229.
[2] Sanche L 2005 *Eur. Phys. J. D* **35** 367 (Review).
[3] Boudaiffa B, Cloutier P, Hunting D, Huels M A and L. Sanche 2000 *Science* **287** 1658.
[4] Martin F, Burrow P D, Cai Z, Cloutier P, Hunting D and Sanche L 2004 *Phys. Rev. Lett.* **93** 068101.
[5] Hanel G, Gstir B, Denifl P S, Scheier P, Farizon B, Farizon M, Illenberger E, Märk T D 2003 *Phys. Rev. Lett.* **90** 188104.
[6] Ptasinska S, Denifl S, Grill V, Scheier P, Märk T D, Gohlke S, Huels M A and Illenberger E 2005 *Angew. Chem. Int. Ed.* **44** 1647.
[7] Abdoul-Carime H, Gohlke S and Illenberger E 2004 *Phys. Rev. Lett.* **92** 168103.
[8] Ptasinska S, Denifl S, Scheier P, Illenberger E and Märk T D 2005 *Angew. Chem. Int. Ed.* **44** 6941.
[9] Abouaf R and Dunet H 2005 *Eur. Phys. J. D* **35** 405.
[10] Scheer A M, Aflatooni K, Gallup G A and Burrow P D 2004 *Phys. Rev. Lett.* **92** 068102.
[11] Ray S G, Daube S S and Naaman R 2005 *PNAS* **102** 15.
[12] Solomun T, Hultschig H and Illenberger E 2005 *Eur. Phys. J. D* **35** 437.
[13] Berdys J, Anusiewicz I, Skurski P and Simons J 2004 *J. Am. Chem. Soc.* **126** 6441.
[14] Simons J 2006 *Acc. Chem Res.* **39** 772-9.
[15] Ptasinska S, Denifl S, Gohlke S, Scheier P, Illenberger E and Märk T 2006 *Angew. Chem. Int. Ed.* **45** 1893-96.
[16] Dabkowska I, Rak J and Gutowski M 2005 *Europ.Phys.J. D* **35** 429-35.
[17] Kumar A, Sevilla M D 2007 *J. Phys. Chem. B* **111** 5464-74.
[18] Bald I, Kopyra J and Illenberger E 2006 *Angew. Chem. Int. Ed.* **45** 4851.
[19] Bald I, Kopyra J, Dabkowska I, Antonsson E and Illenberger E 2007 *J. Chem. Phys.* **126** 074308.
[20] König C, Kopyra J, Bald I and Illenberger E 2006 *Phys. Rev. Lett.* **97** 018105.
[21] Balog R, Langer J, Gohlke S, Stano M, Abdoul-Carime H and Illenberger E 2004 *Int. J. Mass Spectrom.* **233** 267 (Review).
[22] Stamatovic A and Schulz G J 1970 *Rev. Sci. Instr.* **41** 423.
[23] Stano M, Flosadottir H D and Ingolfsson O 2006 *Rap. Comm. Mass Spectrom.* **20** 3498.
[24] NIST Chemistry webbook: http://webbook.nist.gov/chemistry.
[25] Bald I and Illenberger E, in preparation.
[26] Knochenmuss R and Zenobi R 2003 *Chem. Rev.* **103** 441-52.
[27] Frankevich V E, Zhang J, Friess S D, Dashtiev M and Zenobi R 2003 *Anal. Chem.* **75** 6063-67.
[28] Baccarelli I, Gianturco F A, Grandi A, Lucchese R R, Sanna N, Bald I, Kopyra J and Illenberger E 2007 *J. Am. Chem. Soc.* **129** 6269-77.
[29] Hotop H, Ruf M-W, Allan M and Fabrikant I I 2003 *Adv. At. Mol. Opt. Phys.* **49** 85.
[30] Sommerfeld T 2007 *J. Chem. Phys.* **126** 124301.
[31] Sommerfeld T 2005 *J. Phys. Conf. Ser.* **4** 245.
[32] Sulzer P, Ptasinska S, Zappa F, Mielewska B, Milosavljevic A R, Scheier P, Märk T D, Bald I, Gohlke S, Huels M A and Illenberger E 2006 *J. Chem. Phys.* **125** 044304.

Journal of Physics: Conference Series **115** (2008) 012009 doi:10.1088/1742-6596/115/1/012009

Molecular mechanisms for electron-induced damage to biocomponents: cross sections and temporary anionic states for monosaccharides

I. Baccarelli[1], N. Sanna[1], F. A. Gianturco[2], F. Sebastianelli[2]

[1]CASPUR, Supercomputing Consortium, via dei Tizii 6, 00185 Rome, Italy

[2] Department of Chemistry and CNISM, University of Rome La Sapienza, Piazzale A. Moro 5, 00185 Rome, Italy

E-mail: i.baccarelli@caspur.it

Abstract. The electron-molecule collision process for β-D-ribose and β-D-deoxyribose in gas-phase is here characterized in the range 0-20 eV. The description of the ensuing dissociative electron attachment in terms of the Transient Negative Ions (TNIs) is complemented by new calculations on the integral cross sections which show significative differences between ribose and deoxyribose in their furanosic, 5-member-ring conformers. Such finding could suggest a similar response of RNA versus DNA to secondary electrons in the condensed phase with interesting implications for the radiation damage research field.

1. Introduction

There is a variety of processes initiated by the primary radiation impinging on DNA which can induce serious genetic effects, such as mutation [1]. At energies below the DNA ionization threshold the main mechanism responsible for single- and double-strand-breaks (SSBs and DSBs) has been linked to the action of low energy electrons (LEEs) [2, 3, 4] which have been proven to be the most abundant among the secondary species produced by the primary radiation [5]. The essential intermediates between the initial electron collision and the final molecular break-up have been identified as the Transient Negative Ions (TNIs), i.e. metastable states occurring in the scattering process which originate from the temporary trapping of the incident electron in the molecular potential field. The TNIs show a strong 'local' character in the sense that they can essentially be associated to quantum resonances of the single DNA basic constituents (see, e.g. [6, 7]). The characterization of these anionic intermediates is crucial in order to gather critical data for the analysis of the biological consequences of the electron interaction with the complex structures of the living tissues.

The present contribution reviews our most recent advances in the study of electron induced molecular damage as applied to two monosaccharides, the 2-deoxy-D-ribose and the D-ribose. Such basic constituents occupy a central position in the chemical structure of, respectively, the deoxyribonucleic acid (DNA) and the ribonucleic acid (RNA) by binding the phosphate groups into the formation of the well-known backbone structures which provide a link between neighboring nucleotides bases of the single strand of RNA/DNA. We will also complement the analysis of the TNIs arising in the electron-sugar collision by presenting our new results on the corresponding elastic cross sections, which have been obtained by a computational procedure (see Section 2) which solves the same scattering equation by means of a different algorithm.

For both monosaccharides we analysed the two most important conformers [8] identified as the pyranosic, six-member-ring structure and the furanosic, five-member-ring structure. The pyranose is the predominant structure in the gas-phase and, hence, the most relevant in order to compare our theoretical findings with the available experimental results. The second conformer, the furanose, might also be present (in a much smaller percentage) in the gas phase according to the purity of the sample powder vaporized in the experimental chamber [9]. Furthermore, the sugars in the backbones of DNA and RNA are in their furanosic structure, so that the investigation of such conformer of the isolated molecules could provide valuable information for the further analysis of the condensed phase.

The work is organized as follows: the next Section illustrates our two computational procedures, Section 3 reviews our results on the resonance states of both monosaccharides, Section 4 discusses our findings on the elastic cross sections and, finally, Section 5 summarizes our conclusions.

2. The Computational Procedure

2.1. Scattering equations in a single center expansion

The approach we follow to model the electron-molecule interaction is based on a single-centre expansion of the molecular and incident electron wavefunctions about the centre of mass of the molecule (further details on our theoretical procedure when applied to biological systems can be found in Ref. [10]). The expansion basis is given by symmetry-adapted angular functions $\{X_{hl}^{p\mu}(\theta,\phi)\}$ determined by the point group of the molecule [11]. Hence, for instance, the one-electron bound orbitals ϕ_k of the target molecule can be expanded as

$$\phi_k^{p\mu}(\mathbf{r}) = \frac{1}{r}\sum_{hl} u_{hl}^{k,p\mu}(r)\, X_{hl}^{p\mu}(\theta,\phi)\,, \tag{1}$$

where the index k labels a specific, multicenter occupied orbital which belongs to one specific irreducible representation (IR) of the point group of the molecule. The indices $(p\mu)$ label a relevant IR p and one of its components μ. The index h labels a specific basis function, for a given partial wave l, used within the μth component of the pth IR. Finally, the radial coefficients are numerically computed on a grid[12, 13]. In the same fashion all the potential functions modelling the electron-molecule interactions will be expanded in a symmetry-adapted angular basis.

In order to perform the expansion in equation (1) one needs to start from the multicentre wavefunction which describes the target molecule and then generate by quadrature each of the u^k_{hl} coefficients. We generate the bound molecular wavefunction by means of a standard quantum chemistry code, like Gaussian [14]; the explicit expression of the quadrature used to compute the u^k_{hl} coefficients for multicentre Gaussian Type Orbitals (GTOs) has been given in [15].

The Born-Oppenheimer (BO) approximation is assumed to hold and the geometry of the molecular target is kept fixed to the optimized equilibrium geometry for the ground electronic state. We further approximate the target electronic wave function at the Static-Exchange (SE) level with the ground electronic state given by a single Hartree-Fock (HF) determinant in the self-consistent field (SCF) model. The post HF correlation and polarization corrections are included by means of model potentials (later referred to as $\hat{V}^{\mathrm{cp}}$ in Eq. 3 and 4) added to the SE description [16, 12].

The total wave function of the "target + e^-"is then represented as an antisymmetrized product of electronic wave functions which depend on the positions of the fixed nuclei

$$\Psi(\mathbf{r}, \mathbf{X}; \mathbf{R}) = \mathscr{A}\{F(\mathbf{r})\psi(\mathbf{X}; \mathbf{R})\}. \tag{2}$$

where $\mathbf{r}$ is the position of the continuum electron, $\mathbf{X}$ represents collectively the coordinates $\mathbf{x}_i$ (i = 1,...,N) of the target electrons, $\mathbf{R}$ corresponds to the set of frozen nuclear coordinates $\mathbf{R}_\gamma$ (γ = 1,...,M) of the target molecule acting as fixed parameters and, finally, $\mathscr{A}$ is the antisymmetrization operator which operates on the coordinates of the target electrons and of the incident electron. Inserting expression 2 into the Schrödinger equation for the electronic motion and integrating over the target electrons degrees of freedom $\mathbf{X} \equiv \{\mathbf{x}_i\}$ leads to the following integro-differential equation (in a.u.)

$$\left[-\frac{1}{2}\nabla^2 - \frac{1}{2}k^2 + \hat{V}^{\mathrm{st}} + \hat{V}^{\mathrm{cp}} +\right] F(\mathbf{r}) = \int \hat{V}^{\mathrm{exch}}(\mathbf{r}, \mathbf{r}')F(\mathbf{r}')d\mathbf{r}' \tag{3}$$

where **r'** generically represents the coordinate $\mathbf{x}_i$ of one of the target electrons, $\hat{V}^{\mathrm{st}}$ defines the static part of the potential in the scattering process (given by the attractive interaction between the scattering electron and the positive nuclei plus the local part of the repulsive interaction with the target electrons), $\hat{V}^{\mathrm{cp}}$ is the model correlation-polarization potential and $\hat{V}^{\mathrm{exch}}$ is the non-local part of the potential.

2.2. Cross section calculation: solution of Volterra equations

If the exchange term is approximated by a local potential we can rewrite the scattering equation in an homogeneous form to obtain the static-model-exchange correlation-polarization (SMECP) approximation for the scattering equation:

$$\left[-\frac{1}{2}\nabla^2 - \frac{1}{2}k^2 + \hat{V}^{\mathrm{st}} + \hat{V}^{\mathrm{cp}} + \hat{V}^{\mathrm{model\ exc}}\right] F(\mathbf{r}) = 0 \tag{4}$$

As in several previous works, we chose to approximate the exchange interaction by an energy-dependent local function of the electron density of the target by employing the

Hara Free Electron Gas Exchange (HFEGE) potential [17, 18]. Indicating by $V(\mathbf{r})$ the sum of the three local potentials $V(\mathbf{r}) = V^{st}(\mathbf{r}) + V^{cp}(\mathbf{r}) + V^{model\ exc}(\mathbf{r})$, the local formulation of the scattering equation (3), in terms of partial waves and on integrating over the angular variables, takes the form

$$\left[\frac{d^2}{dr^2} - \frac{l(l+1)}{r^2} + k^2\right] F_{lh}^{p\mu}(r) = 2\sum_{l'h'} V_{lh,l'h'}^{p\mu}(r) F_{l'h'}^{p\mu}(r), \tag{5}$$

and the potential coupling elements are given as

$$V_{lh,l'h'}(r) = \langle X_{lh}^{p\mu}(\hat{r})|V(\mathbf{r})|X_{l'h'}^{p\mu}(\hat{r})\rangle = \int d\hat{r} X_{lh}^{p\mu}(\hat{r}) V(\mathbf{r}) X_{l'h'}^{p\mu}(\hat{r}). \tag{6}$$

The standard Green's function technique allows us to rewrite the previous differential equations in an integral form

$$F_{lh}^{p\mu}(r) = \delta_{lh}\, j_l(kr) + \sum_{l'h'} \int_0^r dr'\, g_l(r,r') V_{lh,l'h'}(r') F_{l'h'}^{p\mu}(r). \tag{7}$$

The upper limit of the integral on the right-hand side of the previous equation is a variable (integral equations with this property are called Volterra equations) and the integral vanishes for $r' > r$ because the regular Green's function $g_{l_i}(r,r')$ is used

$$g_l(r,r') = \frac{1}{k}\left[\hat{j}_l(kr)\,\hat{n}_l(kr') - \hat{n}_l(kr)\,\hat{j}_l(kr')\right], \tag{8}$$

where $\hat{j}_l(kr)$ and $\hat{n}_l(kr)$ are Riccati-Bessel and Riccati-Neumann functions.

The four structures under study show no symmetry elements, i.e. they belong to the C_1 symmetry group; hence, no reduction of the size of the problem into subgroups corresponding to the appropriate irreducible representations can be carried out. In our calculations we include the contribution of all the partial waves up to l_{max}=30 for the scattering wavefunction (giving a total number of 961 coupled channels) while the potential is expanded up to L_{max}=60 (corresponding to 3721 partial waves). For each scattering energy, from the K matrix obtained after the Volterra integration we derive the transition T matrix and, finally, the integral cross section (ICS) at that collision energy.

2.3. Resonances calculation: the piece-wise diabatic model

An alternative expansion basis to that of Eq.(1) is provided by the eigenfunctions obtained from diagonalizing the angular Hamiltonian at each radius r [19, 20]. These new angular eigenstates are referred to as the adiabatic angular functions (AAFs) $Z_k^{p\mu}(\theta,\phi,r)$ which are distance-dependent, linear combinations of the symmetry-adapted "asymptotic" harmonics mentioned before

$$Z_k^{p\mu}(\theta,\phi,r) = \sum_{lh} X_{lh}^{p\mu}(\theta,\phi) C_{lh,k}(r), \tag{9}$$

where the expansion coefficients are solutions to the matrix eigenvalue equation

$$\sum_{lh} V_{l'h',lh}(r) C_{lh,k}(r) = V_k^{A}(r) C_{l'h',k}(r). \tag{10}$$

Journal of Physics: Conference Series **115** (2008) 012009 doi:10.1088/1742-6596/115/1/012009

The eigenvalues $V_k^A(r)$ of the angular Hamiltonian now form an adiabatic radial potential for each index value k over the selected range of the e^--molecule distances [12, 21].

Solving the scattering equations using these new $V_k^A(r)$ potentials allows the expansion of the scattering wave function in adiabatic angular states to converge more rapidly than the corresponding expansion in angular momentum eigenstates [19]. Thus, the number of coupled radial equations which must be solved with the adiabatic basis set is usually much smaller. Furthermore, the numerical instabilities in the solution of the standard momentum eigenfunction expansion when very large angular momentum values have to be considered (as in the present molecules) are greatly reduced. On the other hand, the non-adiabatic radial coupling introduces additional terms in the radial differential equations for which the standard integration method is not directly applicable.

In order to avoid the non-adiabatic coupling terms we actually employ a piecewise diabatic (PD) representation for the potential [12] where the radial coordinate is divided into a number of regions. In each radial region we average the coupling potential $V_{l'h',lh}(r)$ over r and the resulting averaged potential is diagonalized to yield a set of angular functions $Z_{k,i}^{p\mu}(\theta,\phi)$ for each region i thereby transforming the scattering potential into a new representation in which it is nearly diagonal. The resulting equations are then solved using the full scattering potential in each region with the further approximation of ignoring the off-diagonal couplings within the region (for further details see Ref.[20, 12]).

A narrow and isolated resonance in a scattering process at an energy E_R and width Γ is associated to a pole in the scattering (S) matrix which has been analytically continued into the complex plane at an energy $E = E_R - i\Gamma/2$ [22]. The S matrix is obtained from a scattering wavefunction by finding a solution with the asymptotic form

$$lim_{r\to\infty}\Psi_{lh,l'h'}(r) = h_l^-(kr)\delta_{ll'}\delta_{hh'} - S_{lh,l'h'}h_l^+(kr) \tag{11}$$

where $h_l^+(kr)$ and $h_l^-(kr)$ are the appropriate Hankel functions. Then it is possible to compute directly the S matrix at complex energies by computing the solution using the standard numerical procedure with a complex valued energy and by matching the solution to the Hankel functions with the appropriate complex argument. A resonance energy in the scattering process can be located by finding those energies at which $0 = 1/\mathrm{detS}$.

As already mentioned, the use of adiabatic angular functions enables us to solve for a smaller number of coupled equations so that we can attain a good degree of convergence in a reduced expansion basis. However, the computational weight of the electron scattering problem for the large, low-symmetry biological molecules sets a limit to the dimension of the final expansion basis. Furthermore, the use of adiabatic angular functions implies a simplification in the evaluation of the radial coupling elements. Hence, we expect some possible shifts in the energy location of the complex S-matrix poles with respect to the resonance peaks shown by our computed cross sections. In fact, as illustrated in the previous Section 2.2, the elastic scattering probabilities are found

Journal of Physics: Conference Series **115** (2008) 012009 doi:10.1088/1742-6596/115/1/012009

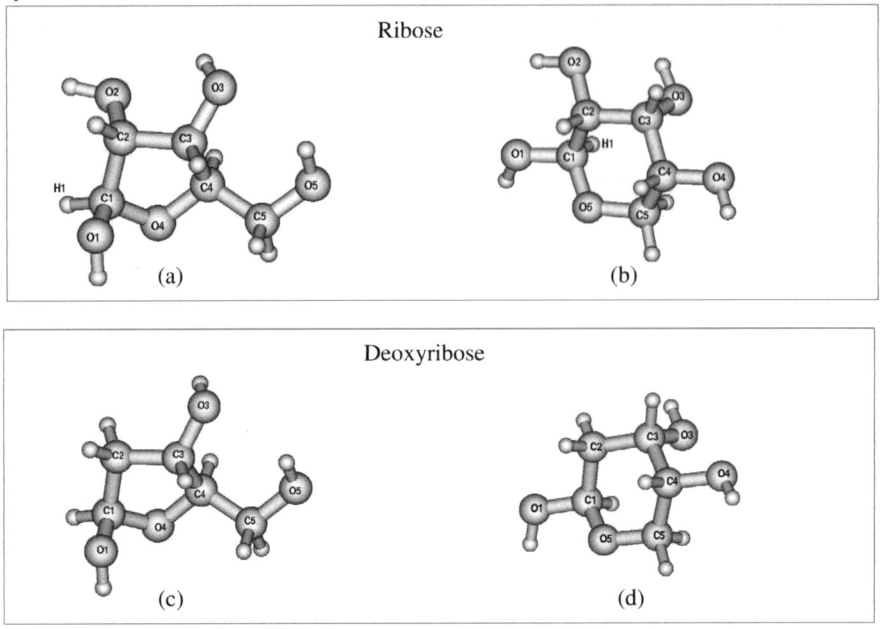

Figure 1. Optimized structure and atoms labelling for the furanose (panel (a)) and the pyranose (panel (b)) of the β-D-Ribose and the corresponding conformers for the β-D-DeoxyRibose (panel (c) and (d)) .

by solving the full scattering problem. For both approaches we need instead to consider possible effects caused by the use of a simplified, local potential and dipole effects at low energies. However, the main focus of the present paper, as well as of our previous work, is to examine in some detail the mechanism and the qualitative characteristics of low-energy one-electron resonances. The analysis requires a model which is simple enough to be applied to large molecules with very few symmetry elements (even none as in the present work) but includes sufficient details of the full scattering problem to accurately reproduce the essential features of the more accurate exact-exchange calculations. Although it is possible that the nonlocality of the exchange potential could provide an additional mechanism for resonant trapping, the good qualitative agreement between results obtained using exact exchange and local model exchange potentials in the resonance analysis (see, e.g. Ref. [23]), gives us confidence on the reliability of the present work. The threshold region is, on the other hand, more critical as for the role of the exact exchange potential on the cross sections, so that in the present work we will limit our discussion of the ICSs to the energy range beyond 2 eV.

3. Resonances and fragmentation channels

We review here our main results concerning electron attachment to isolated ribose [24] and deoxyribose [25] in the collisional energy range 0-20 eV. The four optimized structures for the furanosic and pyranosic structures of the β-D-ribose and the β-D-deoxyribose are reported in Figure 1

Table 1. Position (in eV), width Γ (in eV), associated lifetime (in fs) $\tau = h/\Gamma$ and time t_{e^-} required by a non-attaching electron travelling at the resonance energies to pass through a molecular box of 6 Å for the TNIs found for the β-D-Ribose and the β-D-Deoxyribose in the range 0-20 eV. In italic the poles added with respect to our previous works [24, 25].

Ribofuranose				Ribopyranose			
E (eV)	Γ (eV)	τ (fs)	t_{e^-} (fs)	E (eV)	Γ (eV)	τ (fs)	t_{e^-} (fs)
6.41	0.90	4.60	0.400	7.98	1.59	2.60	0.358
8.46	1.93	2.14	0.348	8.23	2.84	1.46	0.353
9.48	2.41	1.72	0.329	8.67	1.35	3.06	0.344
10.63	2.37	1.75	0.310	9.49	3.14	1.32	0.328
12.40	*2.14*	*1.93*	*0.287*	*11.17*	*2.66*	*1.55*	*0.303*
12.77	*2.49*	*1.66*	*0.283*	*11.46*	*1.85*	*2.24*	*0.299*
				14.09	*2.79*	*1.48*	*0.270*
Deoxyribofuranose				**Deoxyribopyranose**			
E (eV)	Γ (eV)	τ (fs)	t_{e^-} (fs)	E (eV)	Γ (eV)	τ (fs)	t_{e^-} (fs)
7.13	0.92	4.50	0.379	8.56	1.57	2.63	0.346
8.76	2.12	1.95	0.342	8.86	2.71	1.53	0.340
9.79	2.22	1.86	0.323	9.05	3.29	1.26	0.336
9.95	2.68	1.54	0.321	10.10	2.81	1.47	0.318
12.30	2.91	1.42	0.288	*10.76*	*4.23*	*0.98*	*0.308*
12.88	2.44	1.69	0.282	11.22	2.60	1.59	0.302
				11.87	*4.16*	*0.99*	*0.294*
				14.54	*4.15*	*1.00*	*0.265*

In the ribose case [24] we had limited ourselves to characterize the energy range around 8 eV, where new experimental findings were discussed together with our theoretical investigation; in the present work we extend such range up to 20 eV in order to properly compare the previous results to our new calculations for the cross sections. With respect to [25] we also add here three poles with width larger than 4 eV (which we had arbitrarily taken as the upper limit in the selection of acceptable poles) since the associated lifetime is in the same order of magnitude with respect to the other chosen resonances. In any case it is longer than the time needed by a free electron (moving at the energies given by the resonances' locations) to go across a molecular box of 6 Å as can be seen in Table 1 where we report, for all the four structures we studied, the positions, widths, associated lifetimes and corresponding residence time of a 'transient' electron. Hence, in spite of the large widths, the computed poles can be associated to metastable states once the associated lifetimes are compared to the time that a non-attaching electron would spend in the neighborhood of the molecule.

For both monosaccharides all the computed poles fall in the energy range which corresponds to OH^-/O^- detection in the gas-phase experiments reported in [24, 26]. The analysis of the spatial features of the resonance wavefunctions [24, 25] revealed that the excess charge is localized, in different fashions according to the specific TNI under study, on the oxidrilic groups, on the ring oxygen atoms and on some of the carbon atoms of the rings. The simultaneous presence of antibonding nodal planes across selected bonds of the TNIs suggests, at least in the case of direct, single-break mechanism, what fragments would originate from the TNI dissociation. Such an analysis shows indeed for most TNIs a selective electron attachment to well identified oxydrilic groups which can then be released as anionic fragments due to the antibonding character of the related C-OH bond. However, due to the overlap among the observed resonances, the selective fragment release might be eventually observed, upon isotopic substitution, only for the lowest resonance of the furanosic conformer of both monosaccharides (located at 6.41 eV in ribofuranose case and at 7.13 for its deoxygenated analogue, see Table 1).

Considering a possible energy shift in the locations of the S-matrix poles caused by the approximations discussed in Section 2.3, we could also take the lowest-energy resonances as reasonable candidates to be associated with the ring fragmentation observed between 5 and 7 eV, but the complex dynamics associated to multiple ring breaks is largely beyond our level of characterization of the TNIs. Hence, we did not analyse possible ring-breaking mechanisms leading to the heavier fragments experimentally observed around 6 eV for both monosaccharides [24, 26].

Our calculations do not indicate any resonant structures near threshold while the experiments reveal intense signals from threshold and up to about 1.5 eV [26, 27, 28] leading to hydrogen and water molecule abstractions. We might hence suggest that such resonances do not originate from simple dynamical trapping by the shape of the potential and would need a post-HF description of the target molecule.

4. Cross sections

Following the procedures discussed in Section 2.2 we computed the ICS for the electron scattering by the two monosaccharides in their two conformational structures. In order to compare our findings with the only available data found in literature [29], we also repeated the calculations, in the deoxyribofuranose case, at the lower Static Model Exchange (SME) level, hence neglecting the short-range correlation and the long-range polarization contributions to the electron-molecule interaction and approximating the exchange term by a local model potential. We notice that at the SME level the ICS shows a monotonic, structureless decrease with increasing energy (see the insert in the left panel of Fig. 2). The addition of $\hat{V}^{cp}$ does not substantially change the magnitude of the scattering but tunes the shape of the total interaction enabling the trapping of the incident electron at several collision energies in the range 0-20 eV. With respect to the findings reported in [29] our SME cross section for the deoxyribofuranose is globally larger than the SE cross section of [29] and also qualitatively different in the sense that we do not observe the broad minimum found in [29] around 6 eV.

In the left panel of Fig. 2 we compare the ICS for electron scattering by the furanosic conformers of the two monosaccharides. Both compounds show marked resonant characters in their ICS between 5 and 20 eV in the ribose case and between 3 and 20 eV in the deoxyribose case. For the ribofuranose molecule we can distinguish five

Journal of Physics: Conference Series **115** (2008) 012009 doi:10.1088/1742-6596/115/1/012009

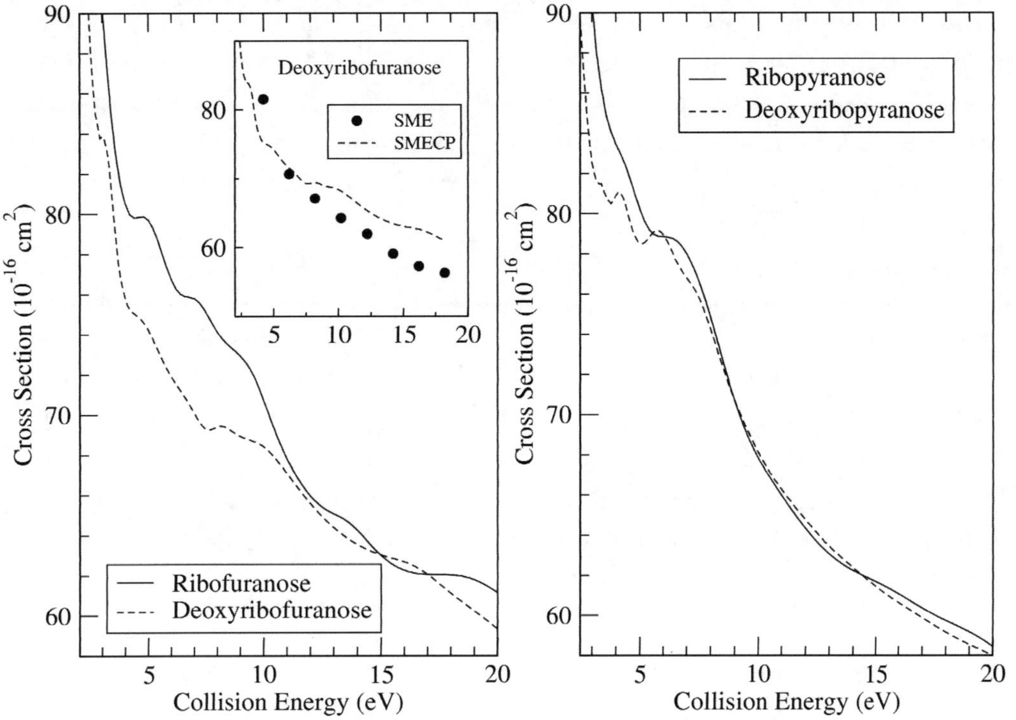

Figure 2. Integral elastic cross sections for electron scattering by β-D-ribose and its 2-deoxy analogue in their furanosic and pyranosic structures. Inset of the left panel: effect of the correlation-polarization model potential on the cross section in the deoxyribofuranose case.

peaks in both cases: at about 5 eV, 7 eV, 9.5 eV, 13.5 eV and 18 eV in the ribofuranose case and at about 3 eV, 5 eV, 8 eV, 10 eV and 16 eV in the deoxyribofuranose case. Excluding for the present the very low energy range, (see Section 2.3) we immediately notice that the two methods we employed to locate the S-matrix poles and to calculate the cross sections, both indicate the 5-15 eV as the interesting region for electron attachment. Furthermore, this is the region where the two monosaccharides differ the most in their ICS, suggesting the ribofuranose to be more reactive towards the electron scattering and attachment in such energy region. Considering that monosaccharides present in the backbones of RNA and DNA are in the furanosic structure, and that we expect some of the features exhibited by the isolated molecules to be carried along in the condensed phase [25], the present finding might provide an interesting piece of information to the wider community working in the radiation damage field.

As for the pyranosic structures, which represent the dominant configurations in the gas-phase, we notice small differences at low energies while the two ICSs are practically indistiguishable over 7 eV.

We are at present carring on a high-level Breit-Wigner analysis of the computed cross sections in order to establish a more quantitative connections between the resonant peaks found in the ICSs and the TNIs already characterized by means of the different computational procedure based on the S-matrix poles calculation in the piece-wise diabatic potential. In such a way we will be able to offer a complete description of the process by 'measuring' the probability for the electron scattering (and attachment) to occur and by analysing the spatial features of the transient anions in order to suggest possible break-up routes.

5. Conclusions

In the present work we complemented our characterization of the electron attachment process by low-energy electrons on the isolated ribose and 2-deoxyribose molecules with the calculations of the elastic scattering integral cross sections in the range 0-20 eV.

The comparison of the computed TNIs and of the peaks shown by the ICSs indicates a substantial agreement in the description of the process at the molecular level. Furthermore, the data on the furanosic, 5-member-ring conformers, show a significative difference which could have important implications in the condensed phase.

Acknowledgments

The financial support of the MUR, Ministery for University and Research, and of the Scientific Committee of the University of Rome "La Sapienza" and the computational support of the CASPUR Supercomputing Consortium are gratefully acknowledged. The FIRB programme is also acknowledged for financial support.

References

[1] C. von Sonntag. *The Chemical basis of Radiation Biology.* Taylor and Francis, London, 1987.
[2] B. Boudaiffa, P. Cloutier, D. Hunting, M. A. Huels, and L. Sanche. Resonant formation of dna strand breaks by low-energy (3 to 20 ev) electrons. *Science*, 287:1658–1660, 2000.
[3] F. Martin, P. D. Burrow, Z. Cai, P. Cloutier, D. Hunting, and L. Sanche. Dna strand breaks induced by 04 ev electrons: The role of shape resonances. *Phys. Rev. Lett.*, 93:068101, 2004.
[4] L. Sanche. Low energy electron-driven damage in biomolecules. *Eur. Phys. J. D*, 35:367, 2005.
[5] V. Cobut, Y. Fongillo, J. P. Patan, T. Goulet, M. J. Fraser, and J.-P. Jay-Gerin. Monte carlo simulation of fast electron and proton tracks in liquid water – i. physical and physicochemical aspects. *Radiat. Phys. Chem.*, 51:229, 1998.
[6] N. A. Richardson, S. S. Wesolowski, and H. F. Schaefer III. Electron affinity of the guanine-cytosine base pair and structural perturbations upon anion formation. *J. Am. Chem. Soc.*, 124:10163–10170, 2002.
[7] N. A. Richardson, S. S. Wesolowski, and H. F. Schaefer III. The adenine-thymine base pair radical anion: Adding an electron results in a major structural change. *J. Phys. Chem. B*, 107:848–853, 2003.
[8] L. P. Guler, Y.-Q. Yu, and H. I. Kenttamaa. n experimental and computational study of the gas-phase structures of five-carbon monosaccharides. *J. Phys. Chem. A*, 106:6754, 2002.
[9] P. Finch. *Carbohydrates: Structures, Synthesis and Dynamics.* Kluwer Academic Publishers, Netherlands, 1999.
[10] I. Baccarelli, F. A. Gianturco, A. Grandi, R. R. Lucchese, and N. Sanna. Electron-driven molecular

processes induced in biological systems by electromagnetic and other ionizing sources. *Adv. Quantum Chem.*, 52:189, 2007.

[11] F. A. Gianturco, D. G. Thompson, and A. Jain. *Computational Methods for Electron-Molecule Collisions.* Plenum, New York, 1995.

[12] R. R. Lucchese and F. A. Gianturco. One-electron resonances in electron scattering from polyatomic molecules. *Int. Rev. Phys. Chem.*, 15:429, 1996.

[13] S. L. Altman and P. Herzig. *Point-Group Theory Tables.* Oxford University Press, Oxford, 1994.

[14] M. J. Frisch, G. W. Trucks, H. B. Schlegel, G. E. Scuseria, M. A. Robb, J. R. Cheeseman, J. A. Montgomery, Jr., T. Vreven, K. N. Kudin, J. C. Burant, J. M. Millam, S. S. Iyengar, J. Tomasi, V. Barone, B. Mennucci, M. Cossi, G. Scalmani, N. Rega, G. A. Petersson, H. Nakatsuji, M. Hada, M. Ehara, K. Toyota, R. Fukuda, J. Hasegawa, M. Ishida, T. Nakajima, Y. Honda, O. Kitao, H. Nakai, M. Klene, X. Li, J. E. Knox, H. P. Hratchian, J. B. Cross, V. Bakken, C. Adamo, J. Jaramillo, R. Gomperts, R. E. Stratmann, O. Yazyev, A. J. Austin, R. Cammi, C. Pomelli, J. W. Ochterski, P. Y. Ayala, K. Morokuma, G. A. Voth, P. Salvador, J. J. Dannenberg, V. G. Zakrzewski, S. Dapprich, A. D. Daniels, M. C. Strain, O. Farkas, D. K. Malick, A. D. Rabuck, K. Raghavachari, J. B. Foresman, J. V. Ortiz, Q. Cui, A. G. Baboul, S. Clifford, J. Cioslowski, B. B. Stefanov, G. Liu, A. Liashenko, P. Piskorz, I. Komaromi, R. L. Martin, D. J. Fox, T. Keith, M. A. Al-Laham, C. Y. Peng, A. Nanayakkara, M. Challacombe, P. M. W. Gill, B. Johnson, W. Chen, M. W. Wong, C. Gonzalez, and J. A. Pople. Gaussian 03, revision c.02, 2004. Gaussian, Inc., Wallingford, CT, 2004.

[15] F.A. Gianturco, J. A. Rodriguez-Ruiz, and N. Sanna. Elastic scattering of electrons by methane molecules. *Phys. Rev. A*, 52:1257, 1995.

[16] F. A. Gianturco, A. Jain, and L. C. Pantano. Electron-methane scattering via a parameter-free model interaction. *J. Phys. B : At. Mol. Phys.*, 20:571, 1987.

[17] S. J. Hara. The scattering of slow electrons by hydrogen molecules. *J. Phys. Soc. Jpn*, 22:710, 1967.

[18] S. Salvini and D. G. Thompson. Exchange approximations in the scattering of slow electrons by polyatomic molecules. *J. Phys. B*, 14:3797, 1981.

[19] M. Le Dourneuf, Vo Ky Lan, and J. M. Launay. Accurate representation of molecular electrons by a few potential-adapted partial waves. *J. Phys. B*, 15:L685–L690, 1982.

[20] F. A. Gianturco and R. R. Lucchese. The elastic scattering of electrons from co2 molecules: Ii. molecular features and spatial symmetries of some resonant states. *J. Phys. B*, 29:3955–3970, 1996.

[21] F. A. Gianturco and R. R. Lucchese. One-particle resonances in low-energy electron scattering from c60. *J. Chem. Phys.*, 111:6769–6786, 1999.

[22] J. R. Taylor. *Scattering theory.* Wiley, New York, 1972.

[23] R. R. Lucchese and F. A. Gianturco. Stort-range effects in resonant electron-molecule scattering from van der waals clusters. *J. Chem. Phys.*, 107:8483–8490, 1997.

[24] I. Baccarelli, F. A. Gianturco, A. Grandi, R. R. Lucchese, N. Sanna, I. Bald, J. Kopyra, and E. Illenberger. Selective bond breaking in beta-d-ribose by gas-phase electron attachment around 8 ev. *J. Am. Chem. Soc.*, 129:6269–6277, 2007.

[25] I. Baccarelli, F. A. Gianturco, A. Grandi, and N. Sanna. Metastable anion fragmentations after electron attachment: deoxyribosic structures from quantum electron dynamics. *Int. J. Quant. Chem.*, 2007. (submitted).

[26] S. Ptasińska, S. Denifl, P. Scheier, and T. D. Märk. Inelastic electron interaction (attachment/ionization) with deoxyribose. *J Chem Phys*, 120(18):8505–8511, 2004.

[27] I. Bald, J. Kopyra, and E. Illenberger. Selective excision of c5 from d-ribose in the gas phase by low-energy electrons (0-1 ev): Implications for the mechanism of dna damage. *Ange. Chem. Int. Ed.*, 45:4851, 2006.

[28] C. König, J. Kopyra, I. Bald, and E. Illenberger. Dissociative electron attachment to phosphoric acid esters: The direct mechanism for single strand breaks in dna. *Phys. Rev. Lett.*, 97:018105,

2006.

[29] C. Winstead and V. McKoy. Low-energy electron scattering by deoxyribose and related molecules. *J. Chem. Phys.*, 125:074302, 2006.

Journal of Physics: Conference Series **115** (2008) 012010 doi:10.1088/1742-6596/115/1/012010

Electron-Impact Ionization of $SiCl_3$ Using An Improved Crossed Fast-Neutral-Beam – Electron-Beam Apparatus

J.M. Mahoney[1,2], M.V. Gutkin[1], V. Tarnovsky[1], and K. Becker[1,3,4]

[1]*Department of Physics and Engineering Physics, Stevens Institute of Technology, Hoboken, NJ 07030, USA*

[2] *Present Address: UV Solutions, Newark, NJ 07103, USA*

[3] *Center for Environmental Systems (CES), Stevens Institute of Technology, Hoboken, NJ 07030, USA*

[4] *Present Address: Polytechnic University, Brooklyn, NY 11201, USA*

[*] Corresponding Author: kbecker@poly.edu

Abstract. The fast-neutral-beam technique is a versatile approach to the determination of absolute cross sections for electron-impact ionization of atoms, stable molecules as well as free radicals and metastable species. A fast neutral beam of the species under study is prepared by charge-transfer neutralization of a mass-selected ion beam and the species are subsequently ionized by an electron beam. Mass- and energy-dispersive selection separates singly from multiply charged ions and parent from fragment ions and allows the determination of partial ionization cross sections. Here we describe some major improvements that were made recently to the fast-beam apparatus that has been used extensively for ionization cross section measurements for the past 15 years in our group. Experiments using well-established ionization cross sections in conjunction with extensive ion trajectory simulations were carried out to test the satisfactory performance of the modified fast-neutral-beam apparatus. We also report absolute partial cross sections for the formation of various singly charged positive ions produced by electron impact on $SiCl_3$ for impact energies from threshold to 200 eV in the modified fast-beam apparatus.

1. Introduction

There is considerable interest in electron-driven collisional interactions, such as electron-impact ionization and dissociative ionization of a molecule, from both a basic science viewpoint and from the viewpoint of data needs for applications. From a theoretical point of view [1], the calculation of electron-impact cross sections for molecular ionization are beyond the capabilities of rigorous quantum mechanical calculations because of the complexity of many molecular targets and the variety of processes contributing to the ionization process. As a consequence, semi-rigorous methods are commonly used [2]. Although experimental ionization study have been carried out since the early part of the 20th century, there are still many atoms, molecules, and free radicals for which no ionization cross section data are available. The reliable determination of partial ionization cross sections are particularly challenging. Mass- and energy-dispersive elements have to be employed to separate singly from multiply charged ions and parent ions from fragment ions. Complete extraction of the product ions from the interaction region of target and electron beams (ion source), transport of the product ions through the mass- and/or energy-dispersive elements (particularly of fragment ions produced with excess kinetic energy) and their detection with 100% efficiency pose serious challenges to the experimentalist.

Märk and co-workers [3] were the first to investigate systematically the ion extraction from the ion source and ion transport through a mass spectrometer from ion source to ion detector and identified discrimination effects as the most serious problem rendering accurate absolute partial ionization cross section measurements a challenging experimental task, particularly in the case of dissociative ionization processes. Detailed investigations of the ion extraction efficiency as a function of excess kinetic energy and of the ion transport efficiency from the ion source to the detector in conjunction with ion trajectory simulations are essential in characterizing and quantifying the ability of a particular apparatus to determine reliable absolute partial ionization cross sections.

Conventional electron-impact ionization apparati use effusive gas jets or heated ovens to produce the target beams for ionization studies. These approaches limit the list of targets to stable atoms and molecules. On the other hand, the fast-neutral-beam method, which was introduced by Cook and Peterson [4], has been shown to be a reliable experimental method for the determination of electron-impact ionization cross sections of free radicals and other short-lived and/or unstable species. Freund and co-workers [5] first described a fast-neutral-beam apparatus used extensively in absolute ionization cross section measurements for many atoms as well as for some molecules and free radicals. This apparatus, in a slightly modified version, was subsequently used in our group in a series of absolute ionization cross section measurements for many molecules and free radicals (see e.g., [6] and references therein to earlier work). Recent technological advances have made it possible to improve the performance of this apparatus significantly (i) by employing position-sensitive ion detection, which allows the determination of the product ion distribution on the face of the detector and (ii) by increasing the electron current density by more than a factor of 10 using a new dispenser-type electron emitter. The impact of these improvements and the expanded capabilities of the modified fast-beam apparatus are described in this paper along with the results of a detailed study of the electron-impact ionization of the $SiCl_3$ free radical.

2. Experimental Apparatus and Performance Verification

2.1 Apparatus Modifications

The fast-neutral-beam apparatus used in the present experiments (and shown in fig. 1) has been described in detail in earlier publications [5,7]. Here we only give a brief summary of two recent

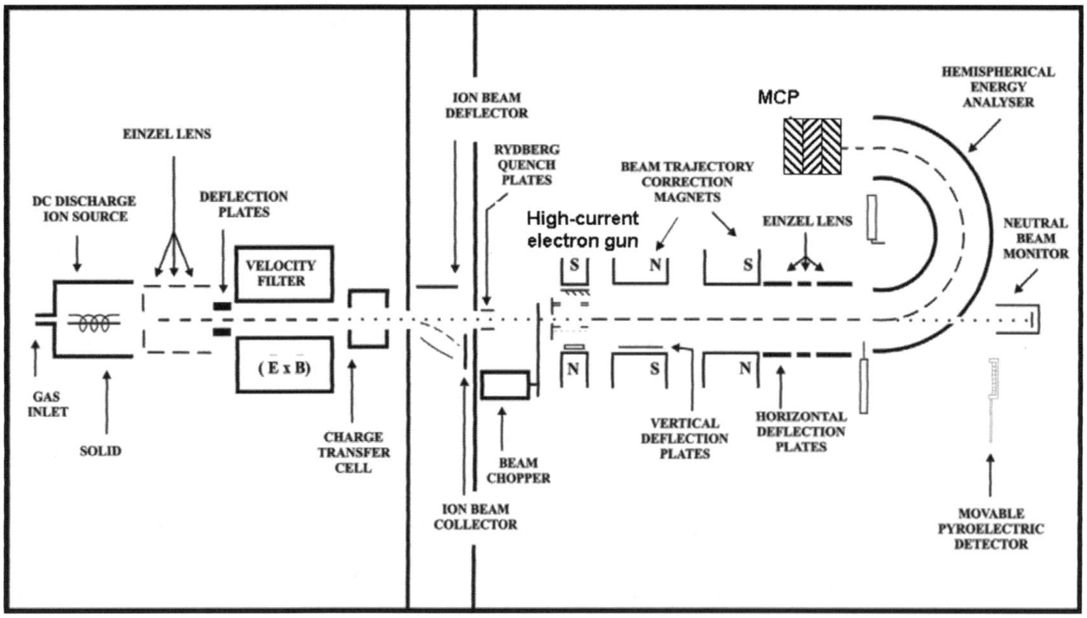

Fig. 1. Schematic diagram of the modified fast-beam apparatus, which employs (i) a positive-sensitive triple-stack multi-channel plate (MCP) detector and (ii) a new dispenser-type high-current electron gun.

Journal of Physics: Conference Series **115** (2008) 012010 doi:10.1088/1742-6596/115/1/012010

modifications to the apparatus and their impact on the performance of the apparatus. First, we installed a new electron gun with a dispenser-type cathode, which consists of a porous tungsten matrix of about 20% porosity as a base, interspersed uniformly with a mixture of barium and calcium aluminate as the electron emitting material. The porous metal matrix acts as a reservoir from which the emitting material can diffuse to the surface, maintain an active layer and provide a low work-function surface for the thermionic emission of electrons. The cathode is activated by indirect heating. The new emitter has the following properties:

- 1-2 A/cm^2 continuous emission current density
- a useful lifetime of more than 10 000 h
- no degradation of the emission current over time
- a minimal evaporation rate
- superior shock and vibration resistance
- high reliability and reproducibility of the operating characteristics

The total beam current produced by the new electron emitter is almost a factor of 10 higher than what was obtained with the previous gun with maximum beam current of well over 2 mA at energies above 50 eV (fig. 2). At beam energies of 10 eV and 4 eV, the total current is still around 20 μA and 5 μA, respectively. In addition, the size (i.e. cross section) of the electron beam produced by the new emitter in the interaction region is about 0.2 cm^2 compared to a beam size of 0.6 cm^2 produced by the old electron gun. Thus, the current density produced by the new emitter of up to 25 mA/cm^2 (for dc operation) exceeds that of the old gun by about a factor of 20.

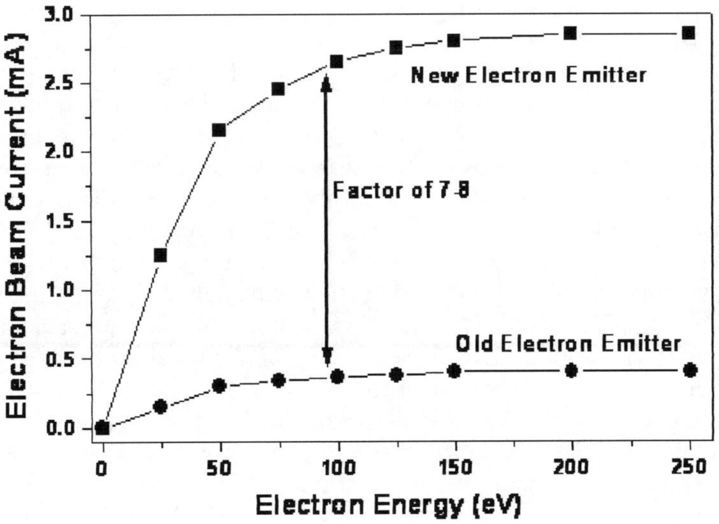

Fig. 2. Comparison of the electron beam current vs. electron energy for the new electron emitter and the old electron emitter.

Furthermore, we also replaced the channel electron multiplier (CEM), which served as ion detector with a position-sensitive, triple multichannel plate (MCP) detector in a Z-stack arrangement for maximum gain (RoentDek model DLD40 MCP detector with a delay-line anode capable of high-resolution 2D-imaging and fast timing for charged particle or photon detection at high rates with multi-hit capability). The new detector allows us to monitor the distribution of the product ions that emerge from the hemispherical analyzer on the face of ion detector. The ability to obtain the product ion distribution in addition to the total ion count is important in situations where fragment ions that are close in their mass-to-charge ratio are formed with broad excess kinetic energy distributions, which causes the individual ion distributions to overlap on the face of the detector. This was the case in our ionization studies of NO, NO_2 and N_2O, where we could not resolve the ion signals corresponding to N^+ and O^+ and only reported a cross section for the combined (N^+ + O^+) formation [5]. The experimental determination of the ion distribution on the detector face in conjunction with SIMION ion trajectory simulations allow us to deconvolute such overlapping ion distributions and obtain individual cross sections.

All other features and components of the original fast-beam apparatus and the experimental procedure to obtain absolute cross sections as described earlier [5,7] remained unchanged. In principle, the fast-beam apparatus affords the capability to measure directly all quantities that determine the absolute cross section. However, here we used the well-established Kr or Ar absolute ionization cross sections to calibrate a pyroelectric crystal. The calibrated crystal, in turn, was then used to determine the flux of the neutral target beam in absolute terms. The typical uncertainty of absolute ionization cross sections determined in the fast-beam apparatus is in the range from ±15 to ±18% [5,7].

2.2 Ion Trajectory Simulations

Extensive ion trajectory simulations using the most recent version of the SIMION charged particle simulation package [9] were carried out to help characterize and quantify the capabilities of the modified fast-beam apparatus. These simulations tracked ions formed in the interaction region of electron beam and neutral beam to the MCP detector. The electron-impact ionization in the interaction region creates product ions, which in the case of molecular targets consist of parent ions and fragment ions produced by dissociative ionization. Parent ions are formed with the same kinetic energy as the incident neutral beam and, in general, no excess kinetic energy is imparted on a parent ion in the ionization process. Fragment ions, on the other hand, are produced with a distribution of excess kinetic energies which may range from near-thermal to more than 10 eV per fragment ion. This results in fragment ions with a much wider kinetic energy spread compared to the parent ions, which, in turn, results in a much more divergent fragment ion beam. Furthermore, the two or more fragments formed in the dissociative ionization process share the initial kinetic energy, so that fragment ions travel with less than the neutral beam energy towards the detector.

In our simulations, we first generate randomized values of the excess kinetic energy of a given fragment ion in a range of values for the excess energy that are known from other experiments or from literature. The energy is converted to a velocity of the fragment, whose direction is randomly selected within a 360° cone. The simulation tracks the path of each ion from the interaction through the electrostatic Einzel lens and the hemispherical analyzer taking into account the displacement of the ions due to the small transverse collimating magnetic field and the effect of the additional magnetic steering field (see fig. 1). Ions that leave the hemispherical analyzer travel an additional few centimeters before they reach the MCP detector, whose entrance has a detection area that is 46 mm in diameter and is held at a negative potential of 3 kV. The simulation ultimately determines the number of ions arriving at the detector and their position on the detector surface and calculates the transmission percentage. If an ion is lost between the interaction region and the detector, it is possible to identify the place where the ion hit a surface or was blocked.

The higher electron beam current from the new emitter and smaller interaction region result in significantly higher signal rates and a much more tightly focused product ion signature on the face of the MCP detector. This is shown in fig. 3, where we simulated the Cl^+ fragment ion signal from the dissociative ionization of $SiCl_4$ (assuming an excess kinetic energy of 3 eV per Cl^+ fragment ion) on the face of the MCP detector for both the old and the new electron gun. In this simulation, the ratio of the signal rates was based on actual values of the beam currents obtained with respectively the old and new electron gun at an impact energy of 70 eV. The effect of the higher current of the new emitter on the signal rate is immediately obvious from the two diagrams depicted in fig. 3. We also note, that the Cl^+ product ion beam obtained with the new emitter is more tightly focused on the face of the MCP detector. Most importantly, the ion collection efficiency with the old emitter was already less than 100% for a 3 eV Cl^+ fragment ion from $SiCl_4$ (note the ions in the left diagram that "miss" the surface area of the MCP), whereas all product ions lie well within the cross sectional area of the MCP when the new electron emitter is employed.

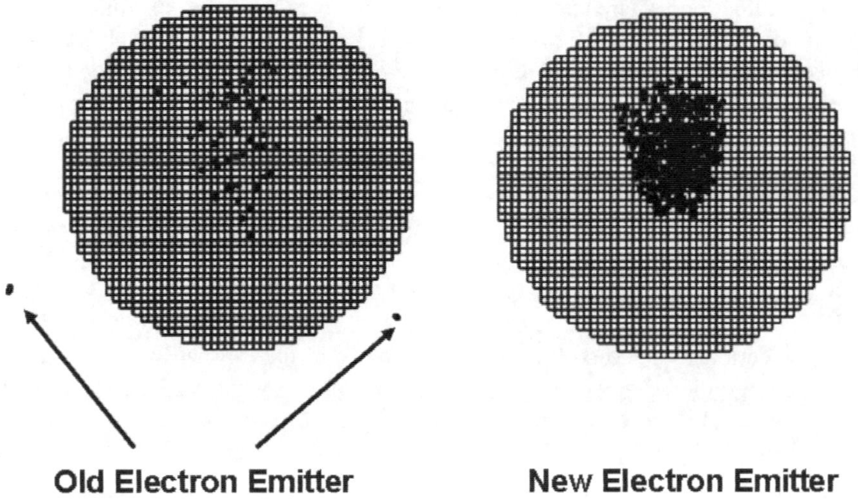

Fig. 3. Comparison of the simulated Cl^+ frag-ment ion beam spot on the detector produced by the dissociative ionization of $SiCl_4$ for a 3 eV excess kinetic energy of the Cl^+ ion. (see text for details).

2.3. Test Measurements

In order to test the performance of the modified fast-beam apparatus, we carried out a series of test measurements using the rare gases (He, Ne, Ar, Kr, and Xe), for which the ionization cross sections (absolute values and cross section shapes) are well known and N_2, where the significant excess kinetic energy of the N^+ fragment formed in the dissociative ionization process poses a serious challenge to the capabilities of the new apparatus.

The absolute ionization cross sections of the rare gases are known with higher accuracy (±4% for Ar, between ±5-±8% for He, Ne, and Kr and ±12% for Xe) than the cross section of any other atom or molecule. The partial rare gas ionization cross sections of Freund and co-workers [5] obtained with the 'old' fast-neutral-beam apparatus are considered among the most reliable data (see e.g. discussion in Ref. [20]). We re-measured the absolute single rare gas ionization cross sections in the new fast-neutral-beam apparatus in the following way. We determined (i) the shape of the Ar^+ single ionization cross section, which has a well-established shape with a very distinct feature around 50 eV and (ii) the ratios of the cross sections for the formation of all singly charged rare gas ions at a fixed energy of 70 eV relative to the Ar ionization cross section. This was done by determining the MCP count rate that corresponds to the established absolute Ar^+ cross section at 70 eV. On the basis of the rare gas ionization cross section ratios at 70 eV as obtained from Ref. [12], we then calculated the expected MCP count rates that correspond to the absolute He^+, Ne^+, Kr^+, and Xe^+ cross sections at 70 eV and measured the actual count rates. The agreement between our measured and calculated count rates was excellent, all values agreeing with each other to well within the margins of error expected on the basis of the accuracy of the underlying absolute cross sections.

Several groups have employed mass-selective techniques to measure partial ionization cross sections for the formation of N_2^+ and N^+ ions from N_2 (see e.g. Peterson [11] and Freund et al. [12]). While the N_2^+ parent ions are formed with thermal or near-thermal kinetic energy, the dissociative ionization leading to N^+ produces fragment ions with a broad excess kinetic energy distribution [1] as evidenced by a N^+ appearance energy of about 30 eV, which is significantly higher than the thermochemical minimum energy of 24.3 eV required to produce N^+ from N_2. We combined SIMION ion trajectory modeling for the formation of N^+ ions from N_2 in the modified fast-beam apparatus with measurements of all partial N_2 ionization cross sections. We simulated the N^+/N_2 ion distribution on the MCP detector with the excess kinetic energy as a free parameter and compared the simulated distribution with measured distributions. A comparison between the measured and simulated ion distributions on the face of the MCP detector reveals (at least qualitative) information about the actual excess kinetic energy distribution and provides a measure of the ion collection efficiency under various operating conditions. Up to an excess kinetic energy of 4.65 eV per N^+ ion, we could demonstrate a 100% N^+ ion collection efficiency. Subsequently, we carried out absolute ionization cross section measurements of N_2 in the modified fast-beam apparatus measuring the N_2^+ and (N^+ + N_2^{++}) partial cross sections as well as the respective appearance energies. The results agreed with the accepted cross section values to better than ±5% and to better than ±0.5 eV for the appearance energies.

3. Results and Discussions

We now present the results of the absolute partial ionization cross section measurements for the $SiCl_3$ free radical from threshold to 200 eV using the modified fast-beam apparatus. Figure 4 shows the measured partial cross sections for the formation of the $SiCl_3^+$ parent ion and all singly charged fragment ions $SiCl_2^+$, $SiCl^+$, Si^+, and Cl^+. These partial ionization cross sections along with the total single $SiCl_3$ ionization cross section are also summarized in Table 1 in the Appendix. Cross sections for the formation of doubly charged ions are not reported here. Similar to what was found in the case of $SiCl_4$ [6], the maximum values of the cross sections for the formation of doubly charged ions from $SiCl_3$ are less than 0.1 x 10^{-20} m^2.

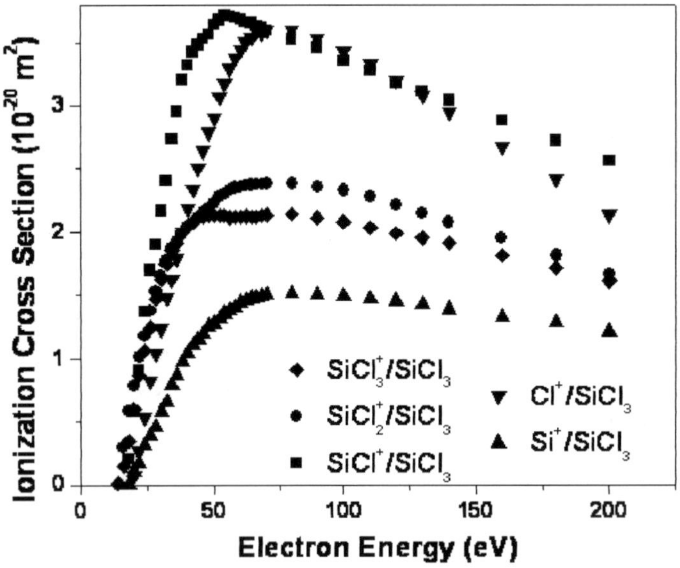

Fig. 4. Absolute partial cross sections for the formation of the singly charged ions $SiCl_3^+$ (diamonds), $SiCl_2^+$ (circles), $SiCl^+$ (squares), Cl^+ (inverted triangles) and Si^+ (triangles) as a function of electron energy from threshold to 200 eV. The absolute cross sections have margins of uncertainty of ±15% (see text), which are not shown.

In terms of their absolute magnitude, the partial ionization cross sections can be grouped into three categories: (1) the $SiCl^+$ and Cl^+ cross sections, which have the largest maximum cross section values (above 3.5 x 10^{-20} m^2), (2) the $SiCl_3^+$ and $SiCl_2^+$ cross sections whose maximum values are slightly above 2 x 10^{-20} m^2, and (3) the Si^+ cross section with a maximum value of about 1.51 x 10^{-20} m^2. The parent $SiCl_3^+$ cross section curve has a threshold of about 12.3 eV and rises to its maximum of 2.14 x 10^{-20} m^2 at an energy of slightly less than 50 eV. The curve then exhibits a slight decline followed by a second maximum of similar magnitude at about 80 eV and subsequently gradually declines to about 1.6 x 10^{-20} m^2 at 200 eV. The $SiCl_2^+$ cross section has a threshold of about 13 eV and rises in a fashion similar to $SiCl_3^+$ cross section, but exhibits only one maximum of 2.4 x 10^{-20} m^2 at about 70 eV. The $SiCl^+$ cross section rises from a threshold of about 14.8 eV to a narrow maximum of 3.7 x 10^{-20} m^2 at about 55 eV and declines to a value of slightly less than 2.6 x 10^{-20} m^2 at 200 eV. The Cl^+ and Si^+ cross sections have both threshold of slightly less than 20 eV. The Cl^+ cross section has a fairly sharp maximum of 3.6 x 10^{-20} m^2 at 70 eV. By contrast, the Si^+ cross section has a broad maximum of 1.5 x 10^{-20} m^2 around 80 eV and a much more gradual decline towards higher impact energies. The shape of the $SiCl_3^+$ cross section with a double-maximum structure is similar to what was observed earlier for selected partial $SiCl_4$ cross sections [6] as well as for some partial cross sections of other Cl-containing molecules, such as $TiCl_4$ [13], and Cl_2 [14]. The low-energy maximum may be indicative of the presence of indirect ionization channels such as autoionization.

Figure 5 shows the experimentally determined total single $SiCl_3$ ionization cross section curve (i.e. the sum of all the partial cross sections for the formation of singly charged ions) in comparison with the calculated total single $SiCl_3$ cross section using the DM formalism. As can be seen, the agreement between measured and calculated total single ionization cross sections is very good in terms of the cross section shape, but less satisfactory in terms of the absolute cross section magnitude. The calculated cross section exceeds the measured cross section at all impact energies above about 30 eV. The discrepancy is largest, about 20% in the region of the cross section maximum. Note that the overall uncertainty in the experimentally determined cross section is about ±17% at 70 eV and is given by the sum of the uncertainties in the relative partial cross sections (added in quadrature) and the uncertainty in the absolute cross section calibration. The level of agreement between calculated and experimentally determined $SiCl_3$ cross section is similar to what was found for other polyatomic molecules (see e.g., Deutsch et al. [2]). We note, however, that by contrast the agreement between calculated and experimentally determined (maximum) cross section for $SiCl_4$ [6] and for $SiCl_2$ and SiCl [15] was better than 5%.

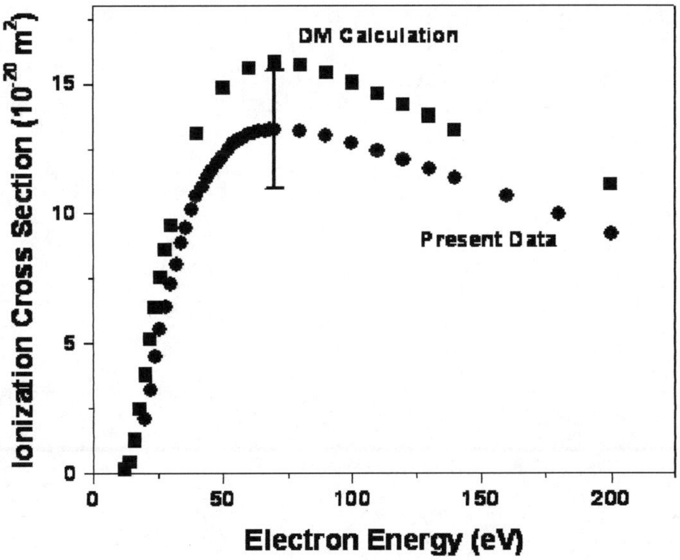

Fig. 5. Absolute total single SiCl3 ionization cross section as a function of electron energy from threshold to 200 eV, present experiment (circles) and calculated cross sections using the DM formalism (squares).

4. Conclusions

Two recent improvements to the fast-beam apparatus that has been used extensively for ionization cross section measurements for the past 15 years in our group have been described, a new high-current electron emitter and a position-sensitive MCP detector. Experiments using well-established ionization cross sections in conjunction with extensive ion trajectory simulations were carried out to test the satisfactory performance of the modified fast-neutral-beam apparatus. We report measured absolute partial cross sections for the formation of all singly charged ions following electron impact on $SiCl_3$ from threshold to 200 eV using the modified apparatus. The maximum cross section values range from 1.5 – 4 x 10^{-20} m^2. Several partial cross section curves show a prominent structure around 30 eV, which may be indicative of the presence of indirect ionization channels such as autoionization. A comparison between the experimentally determined total single $SiCl_3$ cross section and a calculated cross section using the DM formalism shows that the calculated cross section lies systematically above the experimental data. The discrepancy of 20% is similar to what was found previously for other polyatomic molecules, but is larger than the excellent agreement that was found for $SiCl_4$, $SiCl_2$, and SiCl.

Acknowledgments
This work was partially supported by the Chemical Sciences, Geosciences, and Biosciences Division, Office of Basic Energy Sciences, U.S. Department of Energy. The authors would like to thank Mr. Michael Gutkin for his invaluable assistance with some of the measurements described in this article.

References

[1] "Electron Impact Ionization", T.D. Märk and G.H. Dunn (editors), Springer Verlag, Vienna (1985)
[2] H. Deutsch, K. Becker, S. Matt, and T.D. Märk, Int. J. Mass Spectrom. **197** (2000) 37 and references therein to earlier work
[3] H.U. Poll, C. Winkler, and T.D. Märk, Int. J. Mass Spectrom. Ion Proc., **112** (1992) 1
[4] C.J. Cook and J.R. Peterson, Phys. Rev. Lett., **9** (1962) 164
[5] R. C. Wetzel, F. A. Biaocchi, T. R. Hayes and R. S. Freund, Phys. Rev. A **35** (1987) 559
[6] R. Basner, M. Gutkin, J. Mahoney, V. Tarnovsky, H. Deutsch, and K. Becker, J. Chem. Phys. **123** (2005) 05313
[7] V. Tarnovsky and K. Becker, Z. Phys. D **22** (1992) 603
[8] J. Lopez, M. Gutkin, V. Tarnovsky, and K. Becker, Int. J. Mass Spectrom. **225** (2003) 25
[9] Simion 3D, version 7.0, Idaho National Engineering and Environmental Laboratory, Idaho Falls, ID 83415
[10] R. Rejoub, B. G. Lindsay, and R. F. Stebbings, Phys. Rev. A **65** (2002) 042713
[11] J.R. Peterson, Proc. III ICPEAC, London (1963) edited by M. R. C. McDowell, North Holland Publ., Amsterdam (1964), p. 465
[12] R. S. Freund, R. C. Wetzel, and R. J. Shul, Phys. Rev. A **41** (1990) 5861
[13] Basner, M. Schmidt, V. Tarnovsky, H. Deutsch, and K. Becker, Thin Solid Films **374** (2000) 291
[14] R. Basner and K. Becker, New J. Phys. **6** (2004) 118
[15] J. Mahoney, V. Tarnovsky, and K. Becker, Int. J. Mass Spectrom. (2007), in press

Journal of Physics: Conference Series **115** (2008) 012010 doi:10.1088/1742-6596/115/1/012010

APPENDIX

Table 1. Absolute partial and total single electron impact ionization cross sections for $SiCl_3$ as a function of electron energy from threshold to 200 eV.

	Ionization Cross Section (10^{-20} m^2)					
Electron Energy (eV)	$SiCl_3^+$	$SiCl_2^+$	$SiCl^+$	Si^+	Cl^+	Total (single)
12	--	--	--	--	--	--
14	0.01	0.0001	--	--	--	0.0101
16	0.15	0.30	--	--	--	0.45
18	0.35	0.58	0.20	0.004	--	1.13
20	0.60	0.78	0.58	0.10	0.07	2.13
22	0.87	1.01	0.91	0.17	0.27	3.22
24	1.07	1.18	1.37	0.31	0.54	4.472
26	1.24	1.38	1.70	0.39	0.82	5.53
28	1.47	1.53	1.90	0.46	1.04	6.40
30	1.63	1.65	2.17	0.58	1.24	7.27
32	1.75	1.78	2.41	0.66	1.48	8.07
34	1.86	1.87	2.73	0.79	1.62	8.88
36	1.92	1.93	2.95	0.86	1.79	9.46
38	2.00	1.99	3.21	0.98	2.00	10.17
40	2.05	2.04	3.32	1.053	2.19	10.64
42	2.07	2.08	3.431	1.11	2.34	11.03
44	2.11	2.11	3.48	1.15	2.51	11.37
46	2.13	2.16	3.53	1.20	2.64	11.65
48	2.14	2.19	3.56	1.25	2.79	11.94
50	2.13	2.23	3.64	1.28	2.90	12.19
52	2.12	2.28	3.68	1.31	3.07	12.47
54	2.13	2.30	3.71	1.35	3.18	12.68
56	2.12	2.33	3.71	1.38	3.30	12.84
58	2.12	2.35	3.70	1.41	3.38	12.96
60	2.12	2.36	3.68	1.44	3.44	13.05
62	2.13	2.37	3.66	1.46	3.50	13.13
64	2.12	2.38	3.66	1.47	3.54	13.17
66	2.12	2.38	3.63	1.48	3.58	13.19
68	2.13	2.38	3.62	1.49	3.58	13.21
70	2.14	2.39	3.60	1.50	3.60	13.22
80	2.14	2.39	3.53	1.51	3.59	13.16
90	2.11	2.36	3.46	1.50	3.53	12.97
100	2.08	2.33	3.35	1.49	3.43	12.69
110	2.03	2.28	3.28	1.47	3.33	12.40
120	1.99	2.21	3.19	1.45	3.19	12.04
130	1.95	2.15	3.11	1.43	3.07	11.71
140	1.91	2.08	3.04	1.39	2.94	11.36
160	1.82	1.95	2.88	1.33	2.67	10.65
180	1.71	1.81	2.72	1.28	2.41	9.95
200	1.61	1.67	2.56	1.21	2.14	9.19

Multi-temperature moment theory of Ag^+ ion motion and reaction with D_2 in ion traps

Larry A. Viehland[1] and Douglas E. Goeringer[2]
[1]Division of Science, Chatham University, Pittsburgh, PA 15232-2813, USA
[2]Chemical Sciences Division, Oak Ridge National Laboratory, Oak Ridge, TN 37831-6131, USA

E-mail: viehland@chatham.edu

Abstract. A theoretical calculation is reported of the time and space dependence of the reaction rate coefficient in a stretched quadrupole ion trap. The system consists of Ag^+ ions moving in trace amounts through a dilute gas consisting primarily of He or Ar buffer gas, but with a small amount of a reactive neutral, D_2. These systems were chosen because there are available now accurate potential energy curves for Ag^+ ions interacting with He and Ar gas and because the cross section for Ag^+ reaction with D_2 is known over a wide range of collision energy.

1. Introduction

Moment theories of trace ion motion in traps and similar devices have been presented in a series of recent papers [1-5]. In first approximation, the multi-temperature theory [5] for atomic ion motion in atomic gases leads to two vector equations (equivalently, six scalar equations) that constitute a coupled set of ordinary differential equations. The solutions of these equations give the average ion speeds, $v_u(\mathbf{r},t)$, and the dimensionless collision energies, $\epsilon_u(\mathbf{r},t)$, defined below, along the three Cartesian axes ($u=x, y, z$) chosen to describe the apparatus, each as functions of position, $\mathbf{r}$, and time, t. They therefore describe the ion motion in the apparatus in the absence of ion-neutral reactions. For example, numerical values for the $v_u(\mathbf{r},t)$ can be inserted into the equation of continuity, giving a partial differential equation that can be solved for the ion number density, $n(\mathbf{r},t)$, if we are given information about the initial conditions.

When there is a small amount of reactive neutral atoms or molecules mixed in with a much larger amount of non-reactive (buffer) atoms, the values of $v_u(\mathbf{r},t)$, $\epsilon_u(\mathbf{r},t)$ and $n(\mathbf{r},t)$ can be used [6] to determine the two-body, ion-neutral reaction rate coefficient, $k(\mathbf{r},t)$, at a particular position and time, or averaged over the entire apparatus and one or more cycles of the ac fields in the trap. The first purpose of the present paper is to make theoretical calculations of $v_u(\mathbf{r},t)$, $\epsilon_u(\mathbf{r},t)$ and $n(\mathbf{r},t)$ for Ag^+ ions moving in a stretched ion trap filled with He or Ar gas at 300 K, using the accurately known interaction potentials [7] for the $HeAg^+$ and $ArAg^+$ systems. The second purpose is to combine these values with the accurately known [8] integral cross section for Ag^+ reactions with D_2, thus determining the reaction rate coefficient.

2. Theory of the instrument

We assume that the quadrupole ion trap is cylindrically symmetric along the z axis and that a time-dependent electric potential, $U+V\cos(\Omega_{RF}t)$, is applied to the ring electrode. Here U is the amplitude of the dc component of the field and V is the zero-to-peak amplitude of the ac component that has angular frequency Ω_{RF}. We also assume that the hyperbolic end-cap electrodes are identical and arranged symmetrically, and that a time-dependent electric potential is applied between them in a dipolar fashion, i.e., so that one has $[U_D+V_D\cos(\Omega_D t+\delta_D)]/2$ while the other has a negative value of exactly the same magnitude. Here U_D, V_D and ω_D are the equivalents of U, V and Ω_{RF}, while δ_D is the phase shift that may exist between the two ac components. For this arrangement, the total electric potential can be expressed [9] as a multipole expansion of the form

$$\Phi(x,y,z,t) = [U + V\cos(\Omega_{RF}t)]\sum_{l=0}^{\infty} A_{2l}^{R}\left(\frac{r_{3D}}{r_0}\right)^{2l} P_{2l}\left(\frac{z}{r_{3D}}\right)$$

$$+\frac{1}{2}[U_D + V_D\cos(\omega_D t + \delta_D)]\sum_{l=0}^{\infty} A_{2l+1}^{D}\left(\frac{r_{3D}}{r_0}\right)^{2l+1} P_{2l+1}\left(\frac{z}{r_{3D}}\right), \quad (1)$$

where r_0 is the internal radius of the trap rings, $P_l(x)$ is the Legendre polynomial of order l and argument x, and

$$r_{3D}^2 = x^2 + y^2 + z^2 \quad (2)$$

is the distance in three-dimensional (Cartesian) space. The dimensionless expansion coefficients, A_l, have superscripts R and D to serve as reminders that the terms involve the ring and dipole (endcap) potentials, respectively. Their values have been given through l=6 for many different trap geometries [5,9.13].

It is important to note that eq. (1) depends only upon r and z in cylindrical polar coordinates (r,θ,z), where r is the two-dimensional distance in the x-y plane. We assume that the ion trap is symmetric along the z axis about a point (z=0) midway between the endcaps. Hence we will hereafter use cylindrical polar rather than Cartesian coordinates, and we need be concerned only with the region $0 \leq r \leq \infty$ and $0 \leq z \leq \infty$. In practice, of course, the ions do not reach infinity along either axis; indeed, they are generally much closer to the point where both r and z are zero than to any of the electrodes that enclose them.

Here and in each of the other sections of this paper, it is important to give a careful discussion of the possible errors. We will truncate eq. (1) so that only the first six A_l are non-zero. As explained below, we have repeated our calculations while also setting $A_5=A_6=0$. The differences between the theoretical values of the average reaction rate coefficient are negligible, which leads us to believe that keeping only terms through A_6 should lead to no substantial error. There is one caveat to this conclusion, however. While it is entirely feasible to compute the theoretical values of A_l to as high a value of l as may be desired, there is no easy way to assess how well the real ion trap matches eq. (1) with the values assumed for the first six A_l. Overall, we estimate that the errors due to our use of the truncated eq. (1) are less than 0.9%.

3. First-order moment theory

As shown in our previous work [5], the moment equations for the components of the average ion velocity can be cast into forms similar to those ordinarily employed [14] in trap studies of the motion of a single ion. These ordinary differential equations for the average speeds in the radial (u=r) and axial (u=z) directions in cylindrical polar coordinates are

$$\frac{d}{dt}v_u + \frac{\Omega_{RF}^2}{4}[\hat{a}_u - 2\hat{q}_u\cos(\Omega_{RF}t)]u + \xi_u^{(MT)}v_u$$

$$+\frac{\Omega_{RF}^2}{4}\left[\hat{b}_u - 2\hat{d}_u\cos(\omega_D t + \delta_D)\right]\hat{f}_u = 0, \quad (3)$$

where

$$\hat{f}_r = \frac{zr}{r_0} \tag{4}$$

and

$$\hat{f}_u = r_0, \tag{5}$$

and where, for simplicity, we have not explicitly indicated the position and time dependence of the $v_u(\boldsymbol{r},t)$. The quantities $\hat{a}_u$ and $\hat{q}_u$ are modifications [13] of the usual a_u and q_u that are dimensionless versions of U and V for ion traps. The modifications are made using the A_l, and they account for the stretched or otherwise non-ideal nature of the traps. The quantities $\hat{b}_u$ and $\hat{d}_u$, defined previously [13], indicate how the presence of a dipole potential between the endcaps modifies the differential equations. Finally, the collision frequency for momentum transfer along the u axis, $\xi_u^{(MT)}$, depends upon two effective temperatures, $T_r^{(eff)}$ and $T_z^{(eff)}$, as noted below.

It is convenient to introduce the dimensionless effective temperatures,

$$\epsilon_u = \frac{T_u^{(eff)}(r,t)}{3T}, \tag{6}$$

which are implicitly dependent upon $\boldsymbol{r}$ and t. The factor of 3 is introduced into eq. (6) so that, when the ion temperatures are identical, $2\epsilon_r + \epsilon_z$ becomes equal to the ratio of the average ion energy (independent of direction) to the thermal energy, $3k_BT/2$, where T is the gas temperature and k_B is Boltzmann's constant. Since $\xi_u^{(MT)}$ depends upon $T_r^{(eff)}(\boldsymbol{r},t)$ and $T_z^{(eff)}(\boldsymbol{r},t)$, the ϵ_u can also be considered to be dimensionless collision energies.

The moment equations governing the ϵ_u are [5]

$$\frac{d}{dt}\epsilon_u + \frac{\mu\Omega_{RF}^2}{6k_BT}[\hat{a}_u - 2\hat{q}_u\cos(\Omega_{RF}t)]uv_u + \frac{\mu\Omega_{RF}^2}{6k_BT}[\hat{b}_u - 2\hat{d}_u\cos(\omega_D t + \delta_D)]\hat{f}_u v_u + \frac{2\mu\xi_u^{(MT)}}{M}\left(\epsilon_u - \frac{1}{3} + \frac{M\Phi_u^{(MT)}}{m}\epsilon_u\right) = 0. \tag{7}$$

The dimensionless ratios $\Phi_u^{(MT)}$ provide for energy partitioning [15] in the multi-temperature theory, as described below and in the appendix of a previous paper [5]. Like the collision frequencies, they depend upon $T_r^{(eff)}$ and $T_z^{(eff)}$.

Eqs. (3) and (7) are valid only in the first approximation of the multi-temperature moment theory [5]. Although it is possible to extend the theory to second order or beyond, this task is beyond the scope of this paper. For now it suffices to note that momentum-transfer theory [16] and the first approximation of the three-temperature kinetic theory [17] of gaseous ion transport in drift tubes with electrostatic fields are generally accurate within 10%, but for ion/neutral mass ratios as high as for Ag+ in He, they are accurate within 3% [18]. Hence we tentatively estimate that our use of eqs. (3) and (7) should lead to results accurate within 3% for the calculations with He and within 5% for those with Ar. These potential errors are large enough to encompass the small errors that arise from using 100 steps, covering one full cycle of the AC field, to solve eqs. (3) and (7) by a fourth-order Runge-Kutta method [19].

4. Collision frequencies

The collision frequency for momentum transfer in the multi-temperature theory is [5]

$$\xi_0^{(MT)} = \frac{8N\mu}{3m}\left(\frac{2k_BT}{\pi\mu}\right)^{\frac{1}{2}} < \gamma_u^2 >^{(1)}, \tag{8}$$

where N is the gas number density, μ is the ion-buffer reduced mass, and m is the ion mass. The quantity $\Phi_u^{(MT)}$ is

$$\Phi_u^{(MT)} = \frac{3\epsilon_u<\gamma_u^2>^{(2)} - 2\epsilon_r<\gamma_r^2>^{(2)} - \epsilon_z<\gamma_z^2>^{(2)}}{6\epsilon_u<\gamma_u^2>^{(1)}}. \tag{9}$$

Here the generalized collision integrals, $< \gamma_u^2 >^{(l)}$, are energy averages of the usual [20] transport cross sections, $Q^{(l)}$. When cylindrical polar coordinates are used to take advantage of the cylindrical symmetry of the apparatus, the general expression [5] reduces to

$$< A >^{(l)} = 3\sqrt{3}\int_0^\infty d\gamma_r\,\gamma_r\int_0^\infty d\gamma_z\,exp(-\gamma_r^2 - \gamma_z^2)\,A$$

$$\times\, Q^{(l)}(3k_BT[\gamma_r^2\epsilon_r + \gamma_z^2\epsilon_z])\,[\gamma_r^2\epsilon_r + \gamma_z^2\epsilon_z]^{\frac{1}{2}} \tag{10}$$

for any quantity, A, that is an even function of r and/or z. [This equation also corrects important errors in eq. A.4 of [5], due to neglect of the factor of 3 in eq. (6).]

We have used computer program QVALUES [21, 22] to calculate the transport cross sections with a precision of 0.01% for relative kinetic energies between 10^{-9} and 10 hartree, using highly accurate, *ab initio*, interaction potentials [7] for $HeAg^+$ and $ArAg^+$. It is possible that these potentials may still have room for improvement and it is necessary, moreover, to evaluate the generalized collision integrals by numerical quadrature of the double integrals in eq. (10). Nevertheless, we do not believe that the collision integrals, and hence ξ_u and $\Phi_u^{(MT)}$, are in error by more than 0.1%.

5. Ion number density

Since we obtain numerical rather than analytical values of v_u by solving eq. (3), the equation of continuity [1],

$$\frac{\partial n}{\partial t} + \nabla \cdot n\mathbf{v} = 0, \tag{11}$$

also must be solved numerically to obtain the ion number density at any position in the apparatus and at any time. Since cylindrical symmetry ensures that neither n nor **v** depends upon θ, eq. (11) may be written as

$$\frac{\partial n}{\partial t} + \frac{\partial n v_r}{\partial r} + \frac{\partial n v_z}{\partial z} = 0. \tag{12}$$

We have extended our previous computer program [5] to encompass the solution of eq. (12) by the forward-time, forward-space version of the finite difference method [19]. This requires knowledge of *n(r,z,t)* at *t=0*. Assuming that no special experimental techniques have been implemented, this quantity may be assumed to be normally distributed along both r and z. Thus,

$$n(r,z,0) = \left(\frac{2}{\pi}\right)^{1/2} \frac{n_0}{\sigma_r^2 \sigma_z} exp\left(-\frac{r^2}{2\sigma_r^2} - \frac{z^2}{2\sigma_z^2}\right). \quad (13)$$

The constants in eq. (13) ensure that the total ion number in the entire apparatus is n_0, since

$$\int_0^\infty dr\, r \int_0^\infty dz\, n(r,z,0) = n_0. \quad (14)$$

The accuracy of the numerical solution of eq. (13) depends primarily upon the grid size used to cover the positions in the apparatus. In one set of calculations reported below, we used 11 points to cover r values from 0 to four times the standard deviation, σ_r, assumed for the ion number density along this axis and, similarly, 11 points along z. The results appeared generally to be of fair quality, but there were some points at which unphysical results were achieved, i.e., the number density was negative. The curves were of higher quality when we instead used 26 points along each axis, and the unphysical regions occurred at larger values of r and z, where the curves oscillate around 0. A numerical estimate that these effects cause an error of no more than 4% in the reaction rate coefficient is discussed in the next section.

6. Reaction rate coefficient

The multi-temperature expression for the rate coefficient has been given previously [6]. When there is only one reactive neutral species (R) and when the ion trap has cylindrical symmetry, this equation can be put into a form similar to that above for the non-reactive collisions. Thus,

$$k(r,z,t) = \frac{4}{3}\left(\frac{2k_BT}{\pi\mu_R}\right)^{\frac{1}{2}} < 1 >^{(*)}, \quad (15)$$

where

$$< A >^{(*)} = 3\int_0^\infty d\gamma_r\, \gamma_r \int_0^\infty d\gamma_z \exp(-\gamma_r^2 - \gamma_z^2)\, A$$

$$\times\, Q_R^*(k_BT[\gamma_r^2\epsilon_r^* + \gamma_z^2\epsilon_z^*])\, [\gamma_r^2\epsilon_r^* + \gamma_z^2\epsilon_z^*]^{\frac{1}{2}} \quad (16)$$

and where

$$\epsilon_u^* = \frac{m}{m+M} + \frac{M_R}{m+M_R}\frac{m+M}{M}\left[3\epsilon_u - \frac{m}{m+M}\right]. \quad (17)$$

In these equations, M_R is the mass of the reactive neutral, μ_R is the reduced mass of the ion-reactive neutral pair, and Q_R^* is the energy-dependent, integral cross section for reaction.

Eqs. (15)-(17) allow the reaction rate coefficient at a particular position and time to be calculated from the results obtained by solving eqs. (3) and (7). From these results, it is, in turn, possible to compute the average reaction rate coefficient,

$$\bar{k} = \frac{1}{n_0}\int_0^\infty dr\, r \int_0^\infty dz\, k(r,z,t)\, n(r,z,t). \quad (18)$$

Journal of Physics: Conference Series **115** (2008) 012011 doi:10.1088/1742-6596/115/1/012011

However, values of $n(r,z,t)/n_0$ are obtained at a finite set of discrete values of *r*, *z* and *t*, so it is important when converting the integrals in eq. (18) into sums to multiply by the appropriate volume element, $r\Delta r\Delta z\Delta t$, and to avoid the accumulation of round-off errors.

7. Calculations

We applied the theory above to two situations where Ag^+ ions moved through a non-reactive buffer gas, either helium or argon. In each case, a mole faction of 0.1% D_2 was assumed to be present. The ion trap was assumed to be a Finnigan LCQ™ operating under the conditions given in table 1, where initial conditions and other parameters are also collected. We will discuss in detail only the results obtained when the buffer gas was argon and when 100 values along t and 11 values along *r* and *z* were used.

Table 1: Parameters used in the calculations

Parameters	Helium	Argon
r_0, in mm	0.707	0.707
z_0, in mm	0.783	0.783
A_1^R	0.2151	0.2151
A_2^D	-0.5744	-0.5744
A_3^R	0.0978	0.0978
A_4^D	-0.0044	-0.0044
A_5^R	-0.0225	-0.0225
A_6^D	-0.0021	-0.0021
Ω_{RF} in MHz	0.760	1.000
ω_D in MHz	0.110	0.000
δ_D	0.000	0.000
$a_z = -2a_r$	-0.050	0.000
$q_z = -2q_r$	0.400	0.800
$b_z = -2b_r$	0.000	0.000
$d_z = -2d_r$	0.000459	0.000
Range of *r*, in mm	0-2.84	0-2.84
Number of *r* values	11 or 26	11 or 26
Range of *z*, in mm	0-1.42	0-1.42
Number of *z* values	11 or 26	11 or 26
Range of *t*, in μs	0-1.00	0-1.00
Number of *t* values	101	100
$v_r(r,z,t=0)$, in m/s	152.753	152.753
$v_z(r,z,t=0)$, in m/s	152.753	152.753
$3\epsilon_r(r,z,t=0)$	1.000	1.000
$3\epsilon_z(r,z,t=0)$	1.000	1.000
σ_r , in mm	0.710	0.710
σ_z , in mm	0.355	0.355

Figure 1 shows the average Ag^+ speed along the *z* axis as a function of time, at four values of the position along *z* and for two values of position along *r*. Although the curves and dots do not have identical values, $v_z(r,z,t)$ is essentially independent of *r* but a strong function of *z*. This is consistent with eq. (3) when *u*=*z*, where the differential equation has a term directly proportional to *z* but where *r* can affect the results only by its effect upon the ion temperatures, which in turn affect the collision frequencies. Note from Figure 1 that one cycle of $v_z(r,z,t)$ occurs in the same amount of time that it takes for a full cycle of the ac field.

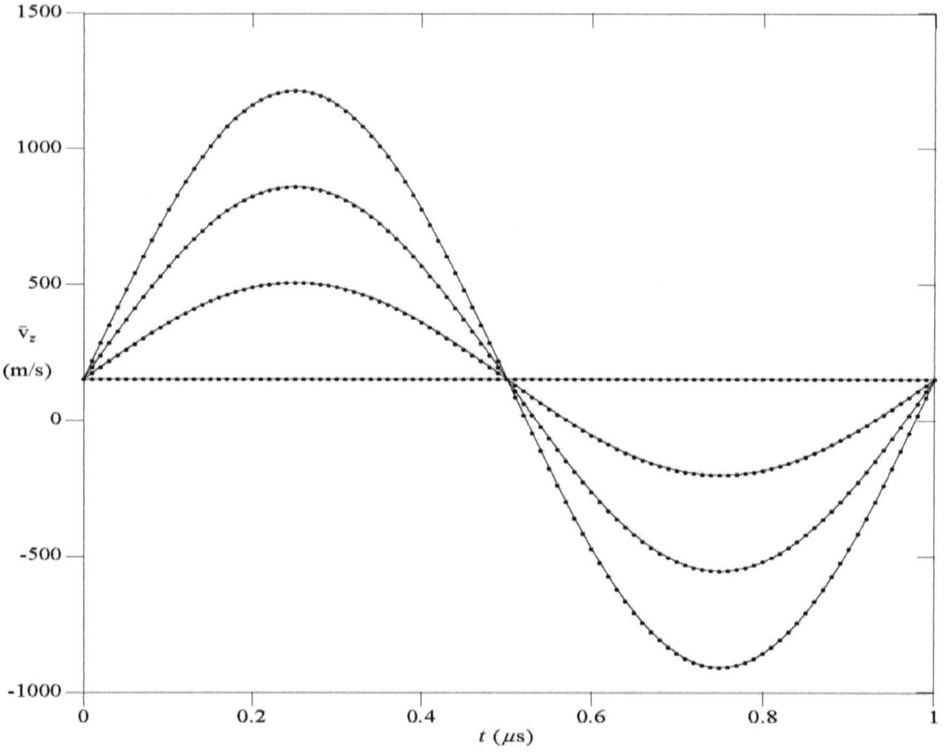

Figure 1: Average Ag^+ speed, in m/s/ along the trap axis (*z* direction) as a function of time, in μs, when the trap contains Ar at 300 K and 0.01 torr. In order of increasing amplitude, the results correspond to *z*=0, 0.142, 0.284 and 0.426 mm from the midpoint of the trap. The curves correspond to values at *r*=0 mm from the trap midpoint, while the dots correspond to *r*=0.852 mm.

Figure 2 shows the average ion speed along the r axis as a function of time. Here we see that $v_r(r,z,t)$ is essentially independent of *z* but a strong function of *r*. Again, this is consistent with eq. (3). It is interesting to note, however, the phase difference between Figures 1 and 2. The ac field along *z* increases during the first part of the cycle, so $v_z(r,z,t)$ is in phase with this field but $v_r(r,z,t)$ is out of phase. This is usual for quadrupole ion traps, and results from the fact that the sum of the dimensionless field terms (in this case, $q_z + 2q_r$) must be zero.

The effective ion temperatures are shown in Figures 3 and 4. As with the average ion speeds, $T_z^{(eff)}(r,z,t)$ is essentially independent of *r* but a strong function of *z*, while $T_r^{(eff)}(r,z,t)$ is essentially independent of *z* but a strong function of *r*. Note, however, that two cycles of these temperatures occur during one cycle of the ac field, a result of the fact that the ion temperature increases when the ion speed increases, whether this increase means more positive or more negative values of the speed. The difference in the magnitude of the two peaks shown in each figure is a result of starting the calculations with initial speeds that are positive along both *r* and *z* (see table 1).

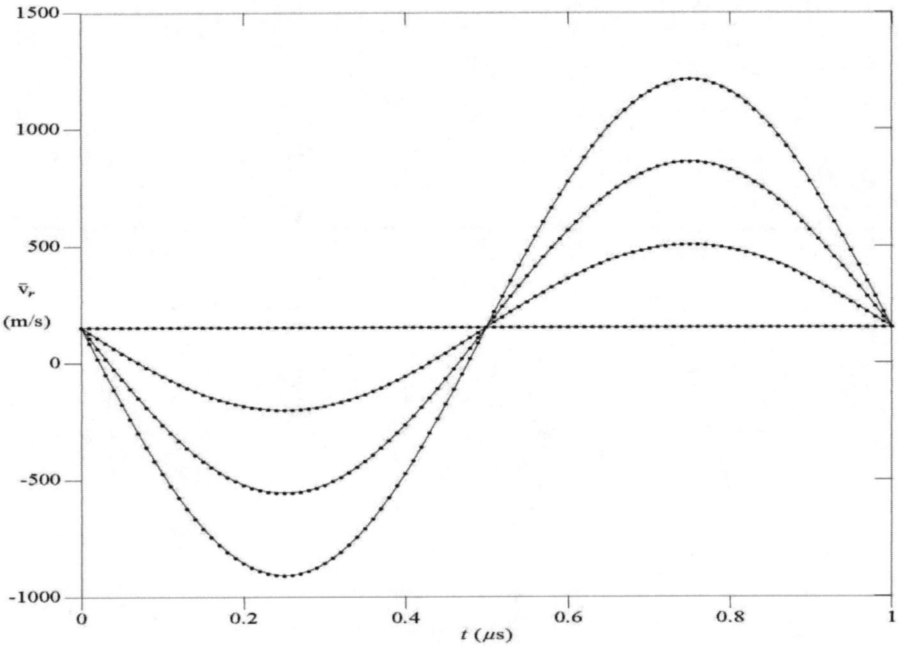

Figure 2: Average Ag^+ speed, in m/s/ along the r direction as a function of time, in μs, when the trap contains Ar at 300 K and 0.01 torr. In order of increasing amplitude, the results correspond to r=0, 0.284, 0.568 and 0.852 mm from the midpoint of the trap. The curves correspond to values at z=0 mm from the trap midpoint, while the dots correspond to r=0.994 mm.

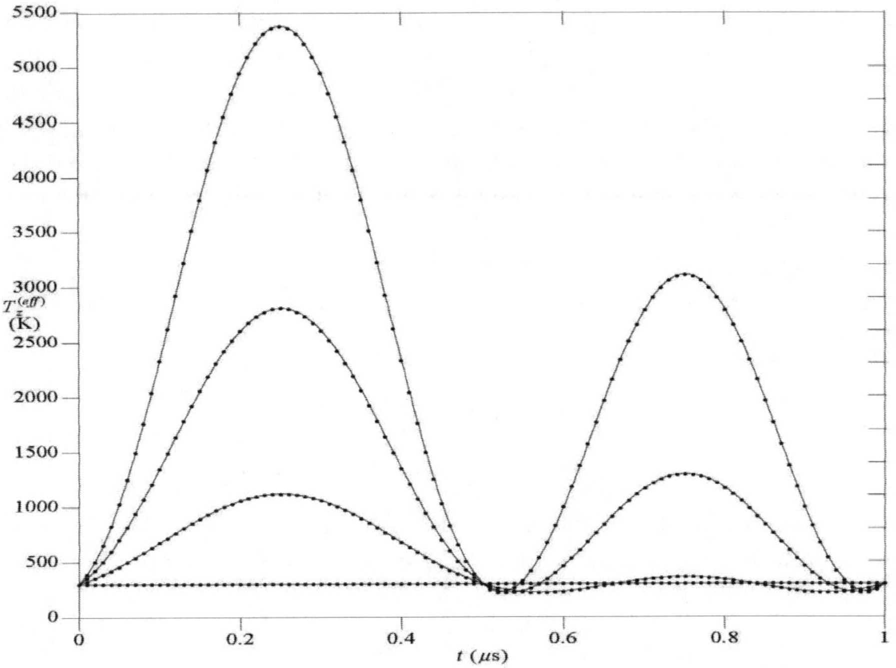

Figure 3: Same as Fig. 1, for the effective ion temperature, in K, along z.

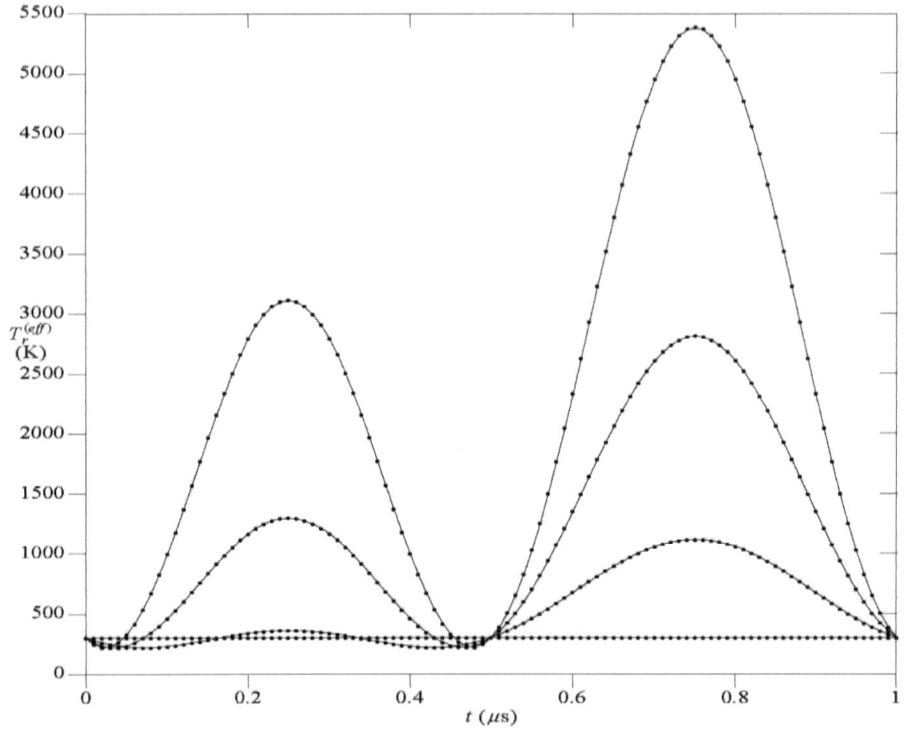

Figure 4: Same as Fig. 2, for the effective ion temperature, in K, along *r*.

The results in Figures 1-4 are consistent with similar results obtained previously [5]. The fact that there is little variation in the average speed or collision energy along either axis in the trap is why the device does indeed trap ions. When a small field is applied in a dipolar fashion to the end-caps, there is a special, "resonance" value of the dipole field that causes ion ejection. What these figures suggest, and what we have shown more conclusively with previous calculations [5], is that a dipole field acts to move ions off the axes, where they encounter large fields due to the ring electrodes, move to higher average speeds and energies, and (when there is a proper match between the dipole and ring fields) are ejected. In short, this is not a true resonance effect.

Completely new results are the effective reaction temperatures shown in Figures 5 and 6. These quantities are related to the ϵ_u^* defined by eq. (17) or, equivalently, they are defined by the equation

$$T_{R,u}^{(eff)}(r,z,t) = \frac{mT+M_R T_u(r,z,t)}{m+M_R}. \qquad (19)$$

The effective reaction temperatures characterize the relative kinetic energies between the ions and reactive neutral molecules. Therefore, Figures 5 and 6 show that few Ag+-D_2 collisions will have energies above the reaction threshold of 1.89 eV (equivalent to 2.19×10^4 K). The reaction rate coefficient will therefore turn out to be quite small.

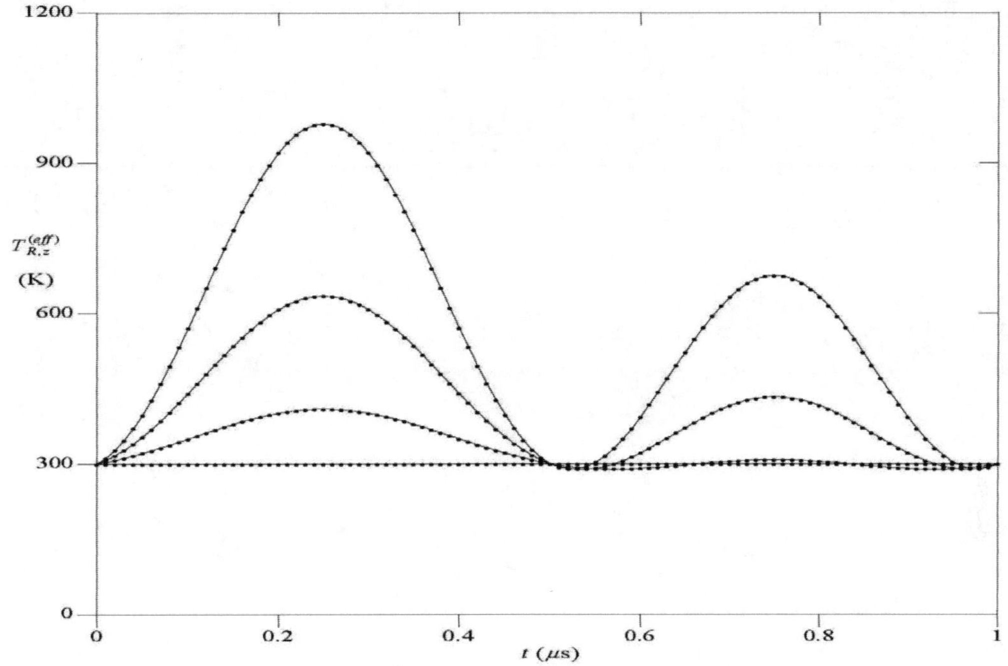

Figure 5: Same as Figs. 1 and 3, for the effective reaction temperature, in K, along z.

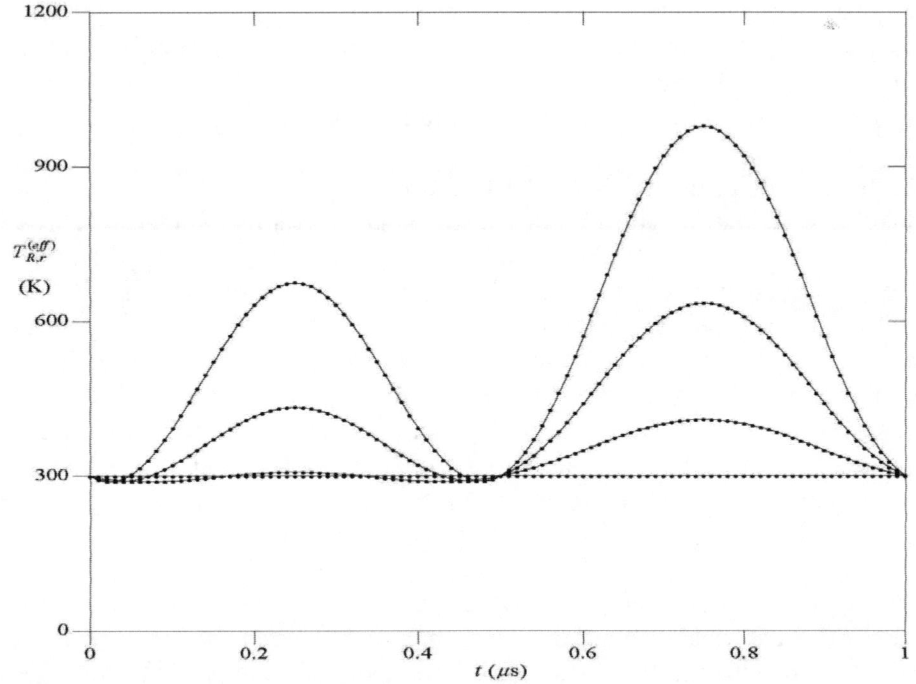

Figure 6: Same as Figs. 2 and 4, for the effective reaction temperature, in K, along r.

Figure 7 shows the ion number density as a function of time and position along *z*, for a set of values of *r*. The values plotted are relative values for points in a particular grid of 11x11 values that characterize the distances along *r* and *z* between 0 and four times the standard deviations given in table 1.

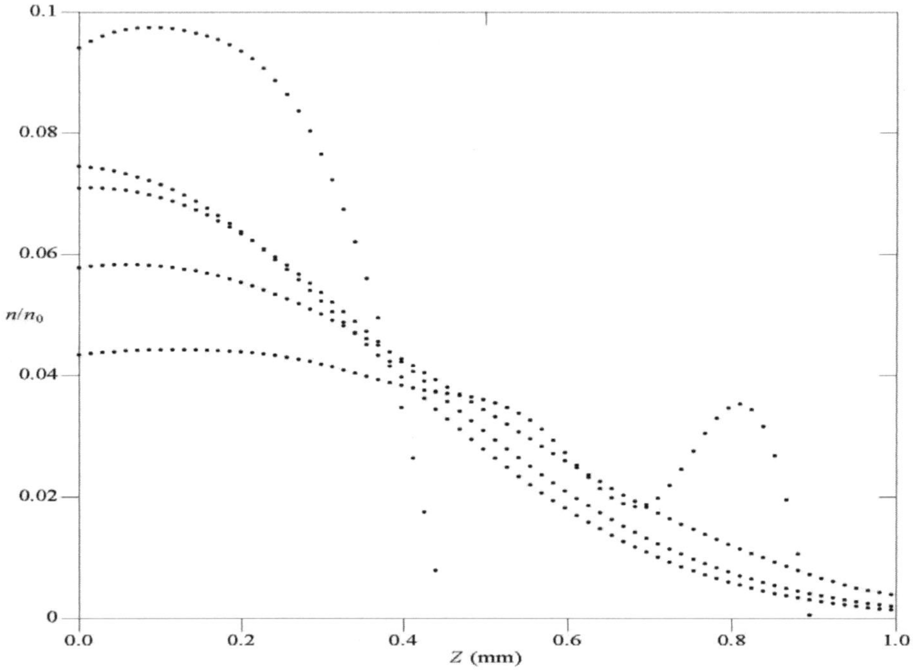

Figure 7: Normalized ion number density, n/n_0, as a function of *z*, in mm, at *r*=0.852 mm. The top set of points corresponds to *t*=0.99 μs. The other sets correspond, in order of decreasing amplitude, to *t*=0, 0.1, 0.2 and 0.3 μs.

We have made calculations similar to those shown in Figure 7 using a 26x26 grid. The n/n_0 values have smaller magnitudes due to the smaller volume element in three-dimensional space that each value represents. More significantly, the unphysical oscillations have smaller amplitudes that occur only at larger values of *z* and *t*. In future work it will be important to choose the number of time steps (100 here) and the size of the grid in *r*-*z* space carefully, using values small enough to keep the calculation time feasible but large enough that the influence of numerical inaccuracies on the final results is negligible.

Note from Figure 7 that the ion density has a maximum at *z*=0, i.e. in the middle of the trap. A similar plot of n/n_0 as a function of t and r but at constant *z* appears to show that the maximum ion number density lies some distance away from the radial axis of the trap. This is artificial and is a result of our use of cylindrical polar coordinates, where the volume element is $r\Delta r\Delta z$.

An important feature of Figure 7 is that it shows that the ion number density at the end of one cycle of the ac field is quite different than at *t*=0, even though it still appears to be a normal distribution. This means that the standard deviations shown in Table 1 are inconsistent with the other parameters, in the sense that they do not describe the steady-state situation corresponding to the other parameters. Future work will involve repeating these calculations until standard deviations are found that that do bring the ion number density after one cycle back to the normal distribution assumed to be present at t=0. Repetitive calculations of this type will be more efficient than solving the moment equations over a large enough number of cycles that steady-state is achieved.

Figures 8 and 9 show the results obtained from eq. (12) for the reaction rate coefficient. Note that the curves and points do not agree, unlike was the case in Figures 1-6. This means that the variation of $k(r,z,t)$ along z is affected by the value of r, and vice versa.

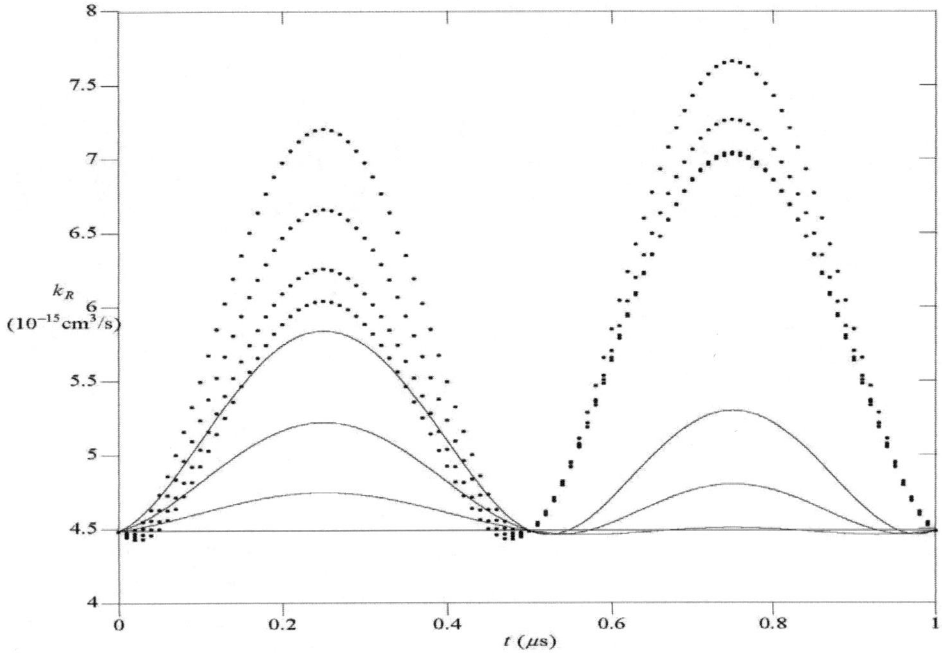

Figure 8: Same as Figs. 1, 3 and 5, for the rate coefficient for the reaction of Ag^+ ions with D_2.

It is a straightforward task to average the results illustrated in Figures 7-9 to obtain the weighted average of $k(r,z,t)$ over all values of r and z and over one cycle of the ac field:

$$\bar{k} = \frac{1}{\tau}\int_0^\tau \left[\frac{1}{n_0}\iint_0^\infty n(r,z,t)k(r,z,t)rdrdz\right]dt. \tag{20}$$

Here τ is one period of the ac field, and the weighting accounts for the different ion number densities at these values of r, z and t.

Our final result is that $\bar{k}$=8.3338×10^{-15} cm^3/s when the buffer gas is argon, when we use 100 steps to cover the time τ, and when we use a 11x11 grid to cover the r-z plane. When we changed to a 26x26 grid but left all other conditions the same, we found that $\bar{k}$=8.0977×10^{-15} cm^3/s; the difference is 2.9%. As indicated previously, we repeated the argon calculations with A_5=A_6=0, in order to assess the effect of possible errors in the theoretical description of the instrument; the resulting values were 8.3317×10^{-15} cm^3/s for a 11x11 grid and 8.0979×10^{-15}

cm^3/s for a 26x26 grid; the differences between these values and those obtained using the values of A_5 and A_6 in table 1 are negligible.

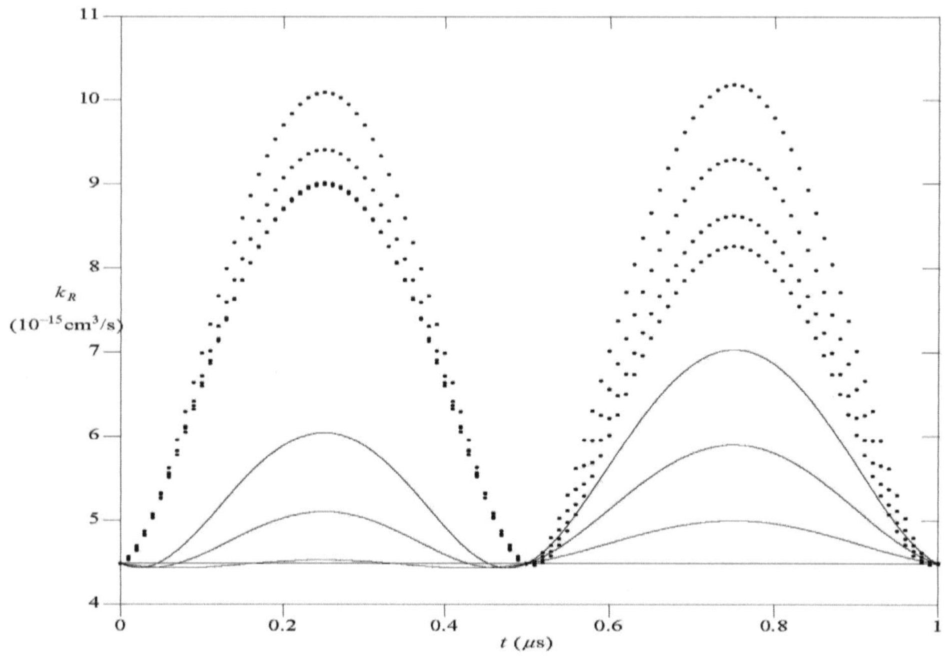

Figure 9: Same as Figs. 2, 4 and 6, for the rate coefficient for the reaction of Ag^+ ions with D_2.

With a helium buffer gas, we found that $\bar{k}$=4.7847×10^{-15} cm^3/s using a grid of 11x11. This value is substantially smaller than in argon, primarily because the value of q_z was much smaller. Somewhat smaller effects are due to the different values of the other parameters in table 1, the much larger ion-neutral mass ratio in helium, and the differences in the ion-neutral interaction potentials. In future work we will explore how variations in the dipole fields (e.g., the values of b_z and d_z) may lead to a resonance that significantly increases the reaction rate coefficient.

8. Discussion

We have calculated the time and space dependences of the reaction rate coefficient for Ag^+ ions reacting with D_2 in a quadrupole ion trap filled with He or Ar. We have also determined the average of the rate coefficient over all positions in the trap and over one full cycle of the ac field. The values of $\bar{k}$ are smaller than the usual lower limit [23] for trap studies, i.e. 1×10^{-13} cm^3/s. This means that reaction times longer than the usual 50 s will need to be used in future measurements of this rate coefficient in a quadrupole ion trap.

The total error in our calculations is estimated to be no more than 7% when the buffer gas is helium, and no more than 9% for argon. This is composed of: 0.9% for possible differences between the real ion trap and the description provided by eq. (1) and the values assumed for the first six A_l; 3% for helium and 5% for argon, due to the use of the first-order moment theory and the use of 100 time steps to cover one full cycle of the ac field; 0.1% for the numerical calculation of the collision frequencies from the ion-neutral interaction potential; and 3% due to the use of a 26x26 grid in *r* and z and the numerical evaluation of eq. (16).

Acknowledgements

This material is based upon work supported by the National Science Foundation under grant no. CHE-0414211. The research of DEG was sponsored by the Office of Basic Energy Sciences, US Department of Energy. Oak Ridge National Laboratory is managed by UT-Batelle, LLC, for the US Department of Energy, under contract no. DE-AC05-00OR22725.

References

[1] Viehland L A and Goeringer D E 2005 *J. Phys. B: At. Mol. Opt. Phys.* **38** 2987
[2] Viehland L A, Kabbe E A and Dixit V V 2005 *J. Phys. B: At. Mol. Opt. Phys.* **38** 4011
[3] Goeringer D E and Viehland L A (2005) *J. Phys. B: At. Mol. Opt. Phys.* **38** 4027
[4] Viehland L A, Danailov D M and Goeringer D E 2006 *J. Phys. B: At. Mol. Opt. Phys.* **39** 3993
[5] Viehland L A, Danailov D M and Goeringer D E 2006 *J. Phys. B: At. Mol. Opt. Phys.* **39** 4015
[6] Viehland L A, Danailov D M and Goeringer D E 2007 *J. Phys. Chem. A* **111** 2820
[7] Yousef A, Shrestha S, Viehland L A, Lee E P F, Gray B R, Ayles V L, Wright T G and Breckenridge W H 2007 *J. Chem. Phys.* **127** 134302
[8] Chen Y-M, Elkind J L and Armentrout P B 1995 *J. Phys. Chem.* **99** 10438
[9] Plass W R 2000 *Int. J. Mass Spectrom.* **202** 175
[10] Beaty E C 1986 *Phys. Rev. A* **33** 3645
[11] Wang Y and Franzen J 1994 *Int. J. Mass Spectrom. Ion Proc.***132** 155
[12] Plass W R 2001, Ph. D. Thesis, Justus-Liebig University, Giessen, Germany
[13] Goeringer D E, Viehland L A and Danailov D M 2006 *J. Am. Soc. Mass Spectrom.* **17** 889
[14] March R E and Todd J F J 1995 "Practical aspects of ion trap mass spectrometry", *Fundamentals of Ion Trap Mass Spectrometry*, vol. 1 (Boca Raton, Florida: CRC Press).
[15] Viehland L A, Mason E A and Whealton J H 1974 *J. Phys. B* **7** 2433
[16] Robson R E 2006 *Introductory Transport Theory for Charged Particles in Gases* (Singapore: World Scientific)
[17] Lin S L, Viehland L A and Mason E A 1979 *Chem. Phys.* **37** 411
[18] Viehland L A and Lin S L 1979 *Chem. Phys.* **43** 135
[19] Asaithambi N S 1995 *Numerical Analysis: Theory and Practice* (Fort Worth: Saunders)
[20] Mason E A and McDaniel E W 1978 *Transport Properties of Ions in Gases* (New York: Wiley) p. 468
[21] Viehland L A 1982 *Chem. Phys.* **70** 149
[22] Viehland L A 1984 *Chem. Phys.* **85** 291
[23] Gronert S 2005 *Mass Spectrom. Rev.* **24** 100

15th International Symposium on Electron Molecule Collisions and Swarms IOP Publishing
Journal of Physics: Conference Series **115** (2008) 012012 doi:10.1088/1742-6596/115/1/012012

Electron-molecule collision cross sections needed for breakdown electric field calculations of hot dissociated SF_6

M. Yousfi[1], Ph. Robin-Jouan[2], Z. Kanzari[1]

[1]Université de Toulouse, Laplace UMR CNRS 5213, UPS, 118 Route de Narbonne 31062 Toulouse Cedex 9, France

[2]AREVA T&D High Voltage Switchgear Research Centre, 130 rue L. BLUM 69611 Villeurbanne, France

Email : yousfi@laplace.univ-tlse.fr

Abstract : The critical electric fields of hot SF_6 are calculated for large temperature and pressure ranges (300 K to 3000 K from 1 bar to several bars). Calculations are based on a multi-term electron Boltzmann equation solution which needs the knowledge of electron-gas collision cross sections for ten SF_6 dissociation products. The collision cross sections are fitted using an electron-swarm unfolding technique. These critical fields are then used to predict the circuit breaker behaviours during the SF_6 recovery phase.

1.Introduction:

Nowadays, SF_6 is widely used in High Voltage gas circuit breaker (GCB) as an arc quenching and an insulating gas. However, due to the size reduction constraints of GCB, the hot gas during its dielectric recovery can be exposed (near the grounded tank or the exhaust tube) to high electrostatic fields leading to undesirable gas breakdown. This therefore leads to a loss of reliability and performance. In order to better understand the undesirable gas breakdown occurrences, it is more particularly necessary to know the breakdown electric fields during the recovery phase of SF_6 which corresponds to a temperature range from about 3000 K down to room temperature. Obviously in such a temperature range, the various SF_6 dissociated by-products (SF_5, SF_4, SF_3, SF_2, SF, S_2, F_2, S and F) can coexist together with SF_6. The proportion of each species directly depends on gas temperature and pressure (see figure 1).

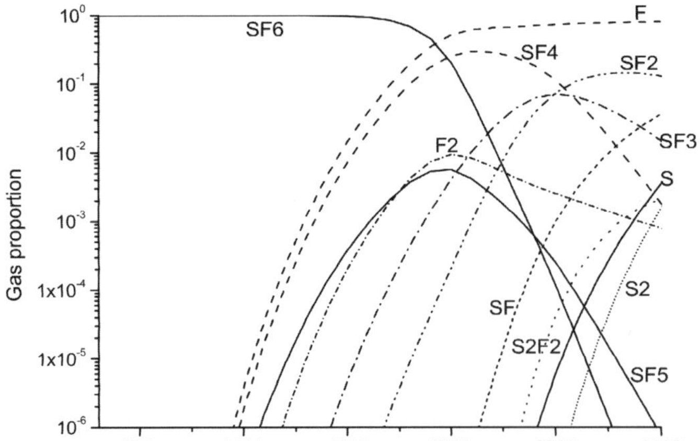

Figure 1 : Relative proportion of SF_6 and SF_6 by-products as a function of gas temperature for a gas pressure of 8 atmospheres (Ref. 1)

Generally, the data on SF_6 breakdown electric field are well known at ambient temperature but in the case of the hot SF_6, they are very sparse in the literature (Ref 2). The hot SF_6 breakdown properties need more particularly a good knowledge of the collisions cross sections for the different electron/SF_6

by-products. In fact, the collision cross sections are rather well known only for electron/SF_6, electron/F_2, electron/F and electron/S systems.

In this paper, a particular emphasis is carried out on the electron/SF_6 by-products collision cross sections needed to determine the critical electric fields of hot SF_6 for large temperature and pressure ranges (300 K to 3000 K and from 1 to several atmospheres). Calculations are based on a muti-term Boltzmann equation solution which needs the knowledge of electron-gas collision cross sections for ten SF_6 dissociation products. The unknown elastic and inelastic collision cross sections are completed from classical relations and fitted using an electron-swarm unfolding technique. Then, these critical fields are used to predict the circuit breaker behaviours during the SF_6 recovery phase.

2. Overview on literature works on hot SF_6 breakdown:

There are a few works in the literature devoted to the measurements of hot SF_6 breakdown (Eliasson and Schade [3], Nagata et al [4], Rothhardt et al [5] and Uchii et al[6]) because it is very difficult to measure the breakdown voltage of a heated gas due to the rapid transient regime occurring during the gas temperature relaxation.

Eliasson and Schade [3] have measured the SF_6 breakdown voltage for a temperature varying from about 1300 K up to 2300 K and a pressure of 2 bar in the wake of cross-arc flow. The gas is heated by an arc shock wave propagating along a channel with a velocity which is measured in order to estimate the gas temperature. The departure from Paschen curve is clearly shown at higher gas temperatures.

Nagata et al [4] have also measured the SF_6 breakdown voltage for a temperature not exceeding 1000 K and for several low gas pressures (50, 100 and 150 Torr). A long graphite tube is used as a gas heating resistance inside the discharge chamber. Due to their low gas temperature (< 1000K), these authors has shown, as expected, that there are no deviation from Paschen law under their experimental works.

Rothhardt et al [5] have used to heat the gas a shock tube technique in which the gas temperature is estimated from the gas flow velocity measurement. Then the breakdown voltage V_b is measured up to 2900 K for a gas pressure of 1 bar. They show several temperature regions in V_b behavior. At a temperature lower than about 800 K, V_b remains equal to its ambient temperature value, then V_b becomes slightly lower up to 2000 K. This is followed by a linear decrease up 2900 K before to become negligible for higher temperatures.

Uchii et al [6] give an interesting work on the hot gas behaviour inside a real circuit breaker. V_b are measured by using small gap electrodes in two different positions in the circuit breaker (at the output of the exhaust tube and near the tank). These measurements are aimed to give cartography of the estimated gas temperature inside the circuit breaker. They conclude that the breakdown of inhomogeneous hot gas is possible as soon as the critical field E_{cr} is reached in any location of the circuit breaker.

As is shown, the literature experimental works are very sparse and concerns only very limited ranges of gas temperature and pressure to allow to correctly predicting the gas circuit breaker behaviours. This is why validated calculations are needed to obtain the breakdown electric field for a large temperature range (300 up to about 3000K) and pressure (1 up 32 atmospheres) corresponding to the recovery phase. These calculations need first of all the knowledge of electron-molecule collision cross sections of the different SF_6 by products present in the hot gas i.e. SF_6, SF_5, SF_4, SF_3, SF_2, SF, F_2, S_2, S and F (see fig. 1). It is important to note the existence of preliminary literature calculations of hot SF_6 breakdown (ref 2, 7 and 8).

Under our gas temperature and pressure the hot breakdown voltage depends mainly on electron-gas interactions.

3. Electron-hot gas interactions:

In the temperature and pressure ranges, ten SF_6 species can have a relative density varying from 10^{-6} up to practically 1 (see fig. 1). As expected, the proportion of SF_6 is dominant at lower temperature, then the proportion of molecular species created from SF_6 dissociation (SF_5, SF_4, SF_3, etc.) after a rise corresponding to their creation phase decreases as a function of the gas temperature to find at higher temperature (i.e. 3000 K) a dominant proportion of F atom species (about 81%) but with a smaller proportion of for example SF_2 (around 12.9%), SF (around 3.7%) and SF_3 (around 1.5%). The interactions between electron and the different gas dissociation products which are able to affect

the electron distribution function (EEDF) are summarized below (for more details, the reader can see Christophorou and Olthoff [9], Christophorou [10], Cliteur et al [8], Miller et al [11], and Yousfi et al [2] with the references given therein these papers). It should be (M being a molecule i.e. SF_x with x = 1 up to 6 or S_2 or F_2 and A being an atom: S or F):
- Elastic collisions: e + M (or A) $\rightarrow$ e + M (or A)
- Vibrational excitations: e + M $\rightarrow$ e + M^v
- Electronic excitations: e + M (or A) $\rightarrow$ e + M^*
- Ionization: e + M (or A) $\rightarrow$ 2e + M^+ (or A^+)
- Attachment: e + M $\rightarrow$ M^- and e + A $\rightarrow$ A^- + hν
- Recombination: e + SF_x^+ $\rightarrow$ SF_x and e + F_y^+ (ou S_y^+)$\rightarrow$ F_y (ou S_y)

Furthermore, it is important to note that ionization can be dissociative as for example in the case of SF_6 where the ionization channel is mainly dissociative leading first to SF_5^+ and in a less degree to SF_3^+, i.e.:

$$e + SF_6 \rightarrow 2e + SF_5^+ + F$$
$$e + SF_6 \rightarrow 2e + SF_3^+ + F_2 + F$$

Attachment processes can also be dissociative or not as for example in the case of SF_6 where several attachment processes are identified:

$$e + SF_6 \rightarrow SF_6^-$$
$$e + SF_6 \rightarrow SF_5^- + F$$
$$e + SF_6 \rightarrow F^- + SF_5$$
$$e + SF_6 \rightarrow SF_4^- + F_2$$
$$e + SF_6 \rightarrow SF_3^- + F_2 + F$$
$$e + SF_6 \rightarrow SF_2^- + 2F_2$$
$$e + SF_6 \rightarrow F_2^- + SF_4$$

Noting that the ion-molecule processes whose the role on the breakdown becomes significant above around 3000K are not discussed in the framework of the present work.

4. Elastic and Inelastic collision cross sections:

Where there are no elastic momentum cross section data available in the literature, we can chose a simplified form of the interaction potential (for example an attractive potential at thermal electron energy i.e. $V(r) = -\alpha/2\, r^4$, α being the atom polarisability and *r* the inter-nuclear distance). Then, by using a specific approximation such as the JWKB one, we can calculate the phase shift δ_l which allows the determination of the elastic momentum transfer cross section $\sigma_m(T)$:

$$\sigma_m(T) = \frac{4\pi}{k^2}\sum_{l=0}^{\infty}(l+1)\sin^2(\delta_{l+1} - \delta_l) \quad (1)$$

k is the wave number of the relative motion proportional to the energy T and $\hbar l$ the angular momentum quantum number.

The excitation cross sections can be obtained from two different analytical formulas concerning separately the allowed and forbidden transition cases. For example, the excitation cross sections for the level *j* in the case of an allowed state can be obtained from a formula consistent with the asymptotic form of Born-Bethe theory (see Ref. [12]):

$$\sigma_{ex\,j}(T) = \frac{q_0 f(1-(W/T)^{\alpha})^{\beta}}{TW} ln(\frac{4TC}{W} + nl) \quad (2)$$

W is the energy lost during the inelastic collision, *f* the oscillator force strength, *C* a correction falloff factor at higher energy and *nl* the base of the natural logarithm while α and β are adjustable parameters.

The total electron-molecule (or atom) ionization cross section can be calculated using the Binary-Encounter-Bethe model (see e.g. Kim et al [13]). This combines the Mott cross section with the high incident energy behaviour of the Bethe cross section. The theory provides a simple analytic formula for the ionization cross section per atomic (or molecular) orbital. The total ionization cross section for

Journal of Physics: Conference Series **115** (2008) 012012 doi:10.1088/1742-6596/115/1/012012

a target is obtained by summing these orbital cross sections. Four orbital constants (the binding energy B, the orbital kinetic energy U, the electron occupation number n_e, and a dipole constant Q) are needed for each orbital, and the first three constants are readily available from the ground-state wave function of the target atom or molecule. The basic formula for the ionization cross section per orbital is:

$$\sigma_{ion}(t) = \frac{S}{t+(u+1)}\left[\frac{\ln t}{2}(1-\frac{1}{t^2})+(1-\frac{1}{t}-\frac{\ln t}{t+1})\right] \quad (3)$$

$t = T/B$, $u = U/B$, $S = q_0\, n_e\, (R/B)^2$, $q_0 = 4\pi a_0^2$ with $a_0 = 0.529$ 18 Å, and $R = 13.6057$ eV, and the dipole constant Q is defined in terms of the continuum dipole oscillator strength df/dW, where W is the kinetic energy of the ejected electron.

Figure 2 shows the ionization collision cross sections by electron impacts on the different molecules and atoms considered in this work (i.e. SF_6, SF_5, SF_4, SF_3, SF_2,SF, F_2, S_2, S and F).

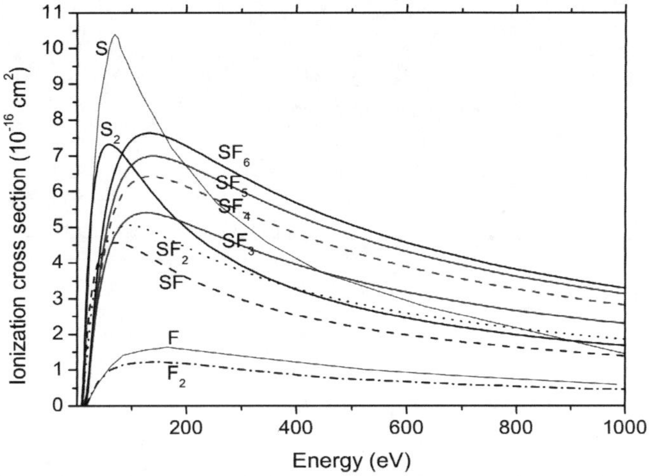

Figure 2: Ionization collision cross sections by electron impacts on SF_6 and its dissociation products

5. EEDF and collision cross section fitting

For a given gas temperature and pressure, we have a specific gas mixture including the different dissociated species shown in figure 1. The breakdown electric field or the critical electric field for such a gas mixture corresponds to the balance between the processes of electron generation and loss. The reaction rates of such processes are calculated from the electron energy distribution function (EEDF). The later needs the a priori knowledge of the collision cross section sets corresponding to the whole species present in the dissociated hot gas as detailed in section 2. EEDF is calculated from a multi-term solution of Boltzmann equation ([14]) which gives more accurate results particularly when the molecular species are abundantly present i.e. for low and intermediate temperature ranges.

The EEDF calculation needs the knowledge of the different sets of electron-gas cross sections. Each set of collision cross sections involve elastic and inelastic processes (see section 4). The sets of collision cross sections for electron/SF_6 (Ref. 14), electron/F_2 (Ref. 15), electron/S and electron/F (Ref.8) are taken from the literature. However, the electron/F and electron/S sets are then completed by adding the attachment cross sections corresponding to radiative attachment (i.e. e + F or S $\rightarrow$ F^- or S^-+ hν) calculated using the principle of detailed balancing (see e.g. Ref. 16) and fitted using the corresponding reaction rates.

In the case where no electron collision cross section sections are available in the literature (i.e. e/SF5, e/SF_4, e/SF_3, e/SF_2 and e/SF), the ionization cross sections are calculated using the binary encounter theory (relation 2 and figure 2) while the excitation cross sections are determined from empirical formulas (as e.g. relation 1) and then fitted to obtain the best coherence between measured and calculated swarm parameters. The method of fitting is also applied to the case of the elastic momentum transfer cross section (see relation 3) also needed for the EEDF calculations.

Journal of Physics: Conference Series **115** (2008) 012012 doi:10.1088/1742-6596/115/1/012012

The obtained collision cross section is therefore fitted to obtain the best coherence with the measured electron swarm parameters. The fitting procedure can be schematically summarized in the flowchart of figure 3.We start from an initial set of cross sections taken partly from the literature and completed with the cross sections whose the shapes obey to the cross section theory (see section 4). Then, the calculated swarm parameters from this initial set are compared to the measured ones. The initial set is modified until the required coherence between measured and calculated swarm parameters is obtained.

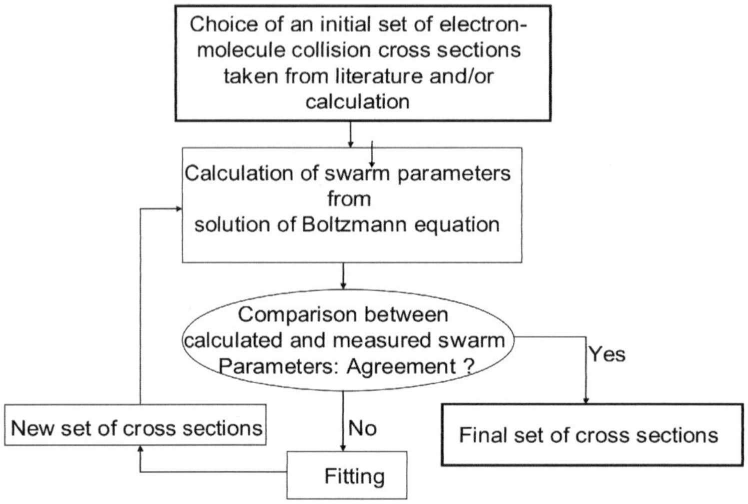

Figure 3: Simplified flow chart of the fitting method of collision cross sections from Boltzmann equation solution

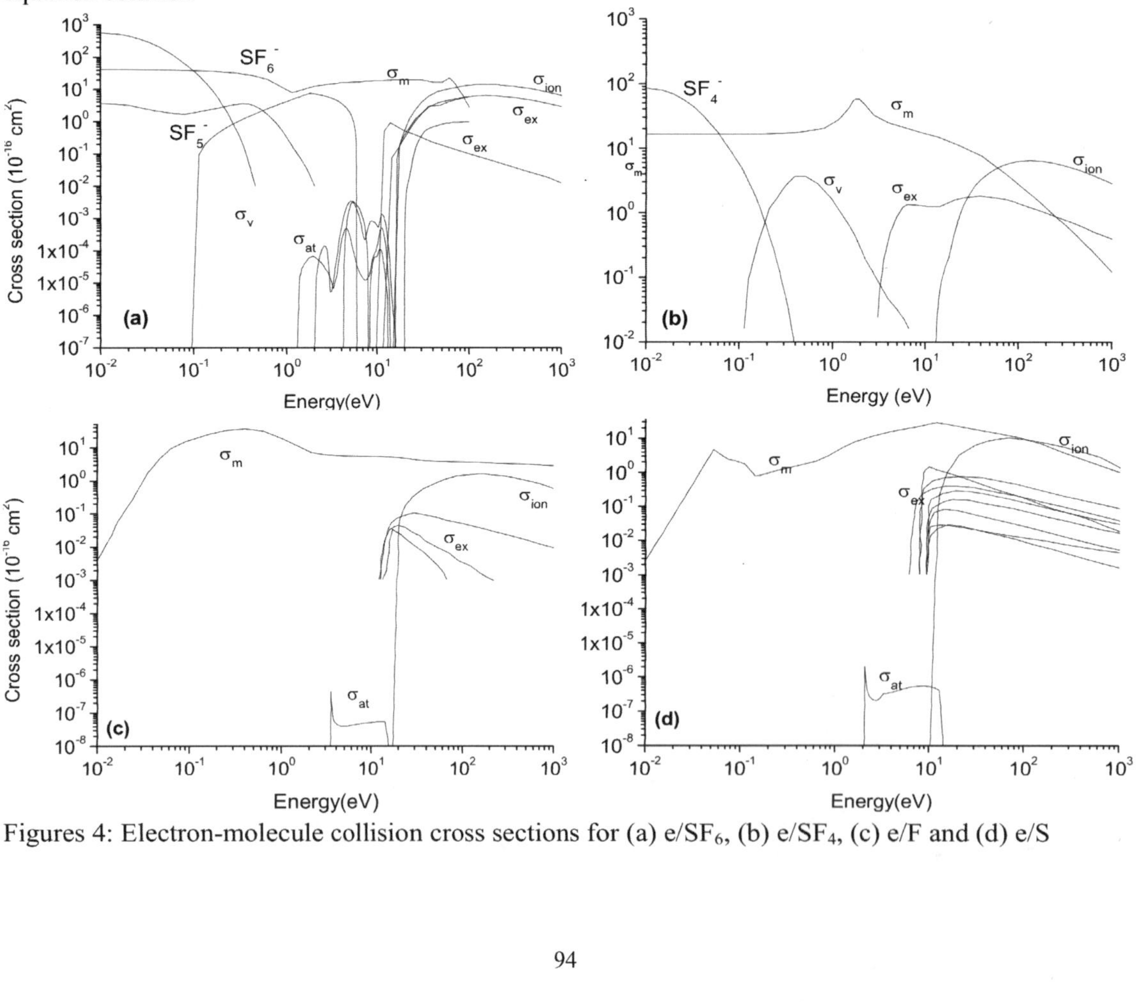

Figures 4: Electron-molecule collision cross sections for (a) e/SF_6, (b) e/SF_4, (c) e/F and (d) e/S

6. Critical electric field as a function of gas pressure and temperature:

The critical electric field is the electric field value corresponding to the balance between the processes of electron generation (ionisation) and loss (attachment).

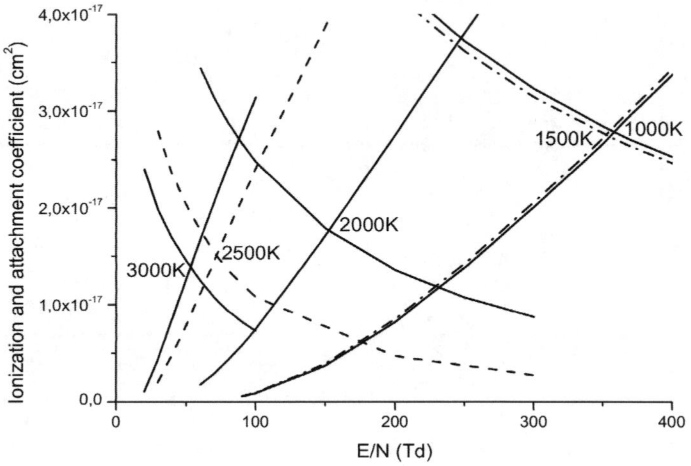

Figure 5: Ionization and attachment coefficients as a function of E/N for different gas temperatures at 8 atmospheres

Figure 5 shows the variation of ionization and attachment coefficients at 8 atmospheres for different gas temperatures. For a given temperature, the critical field Ecr/N corresponds to the E/N value for which ionization and attachment coefficients are equal, i.e. the crossing of the corresponding curves. Thus, for a given pressure when the successive Ecr/N are determined for the whole range of the considered gas temperature we therefore obtain the Ecr/N variation with the gas temperature as shown in next two figures (Figs. 6 and 7).

Figure 6 shows the critical electric field calculated from collision cross sections of hot SF_6 and compared to several data taken from the literature (i.e. Aschwanden [17], Eliasson and E. Schade [3], Cliteur et al [8], Hayashi from Schade [3], Rothhardt et al [5], Yan et al [7].). Present calculations are coherent with the measurement at the ambient temperature ([17]), quite close to Hayashi calculation as soon as the population of F atom becomes dominant, in a qualitative agreement with Cliteur results, close to Eliasson measurements between 1500 K up to 2000 K and in agreement with the Rotthardt estimation between 1900K up to 2300 K.

Such interesting agreement partly validates the sets of collision cross sections considered in this work. This then allows us to extend our critical field calculations to the case of a large pressure range. Figure 7 shows calculated Ecr/N for a quite large gas temperature and gas pressure ranges (i.e. 300 K up to 3000 K and 1 atmosphere up to 32 atmospheres). It can be noted the hot gas keeps roughly the SF_6 dielectric strength as much as the gas dissociation is not significant, this corresponds to the observed first plate on Ecr(T)/N variation which continues up to about 1400 K at 32 atmospheres. We note also that a rapid decrease follows this first plate due to the gas dissociation. This corresponds to a dielectric strength loss of about a factor 3. Then, a slower decrease is observed between 2300K up to 27000 K at 32 atmospheres. This is due to the non negligible presence of SF_4 molecule which shows a maximum in the same temperature range (see figure 1). Finally, at higher temperatures a second plate finishes this slower decrease which roughly corresponds to the dielectric strength of F atom which is the dominant gas composition at least for the lower gas pressure.

Journal of Physics: Conference Series **115** (2008) 012012 doi:10.1088/1742-6596/115/1/012012

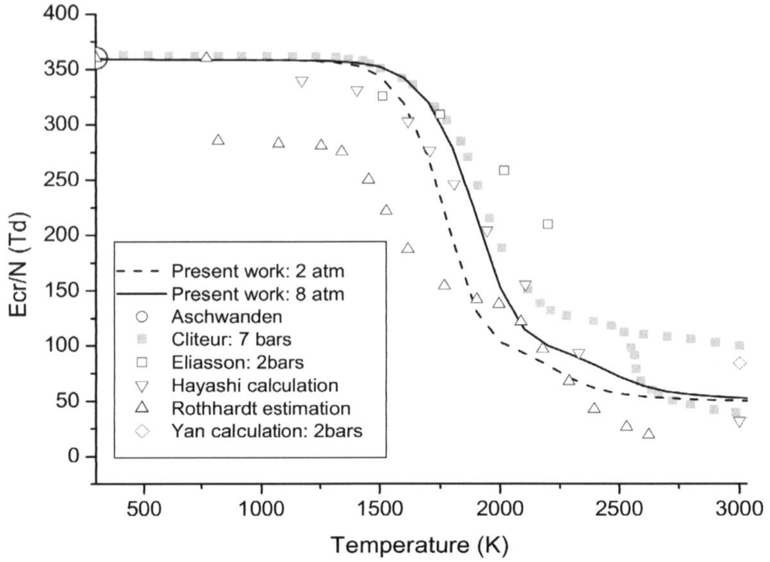

Figure 6: Calculated critical reduced electric fields with comparison to several literature data.

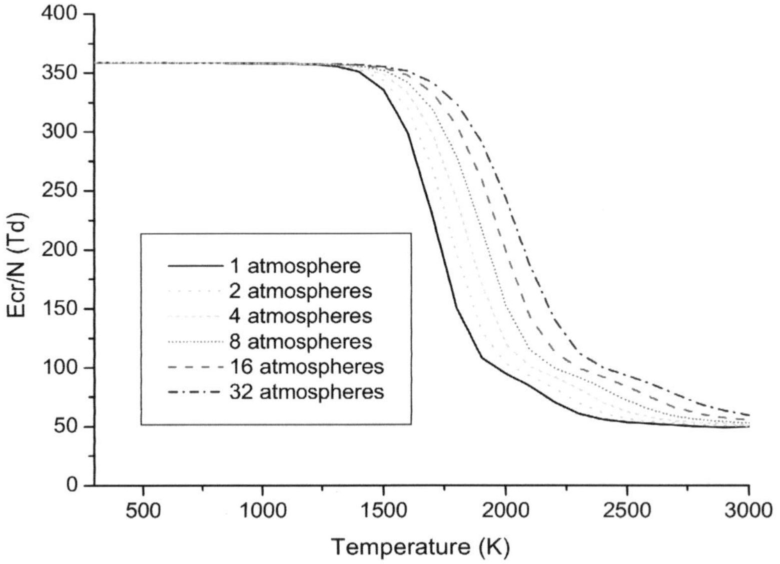

Figure 7: Calculated critical reduced electric field for different gas pressures.

7. Application to circuit-breaker simulations:

As is known, the different processes of the electric arc are linked to many complex physical phenomena: high compressibility, turbulence, electromagnetism, gas ionisation, arc radiation, and wall ablation. Assuming the local thermodynamic equilibrium, we can apply the classical conservation equations for mass, momentum and energy using the Euler's equations [18]. The electric arc introduces in these equations some extra coupled terms: Joule and radiation effects in the energy equation, magnetic forces in the momentum transfer equations. Electric and magnetic fields are required and calculated through the simplified Maxwell's equations. The radiation and erosion are taken into account. Gas ionisation is taken into account through the non-linear variations of the thermodynamic and transport properties of SF_6 with respect to the temperature and pressure.

Journal of Physics: Conference Series **115** (2008) 012012 doi:10.1088/1742-6596/115/1/012012

The resolution is based on a finite volume method and uses a non-structured triangular mesh which is automatically adapted to the movement of the different parts of the circuit breaker and to the temperature, density and electrical gradients.

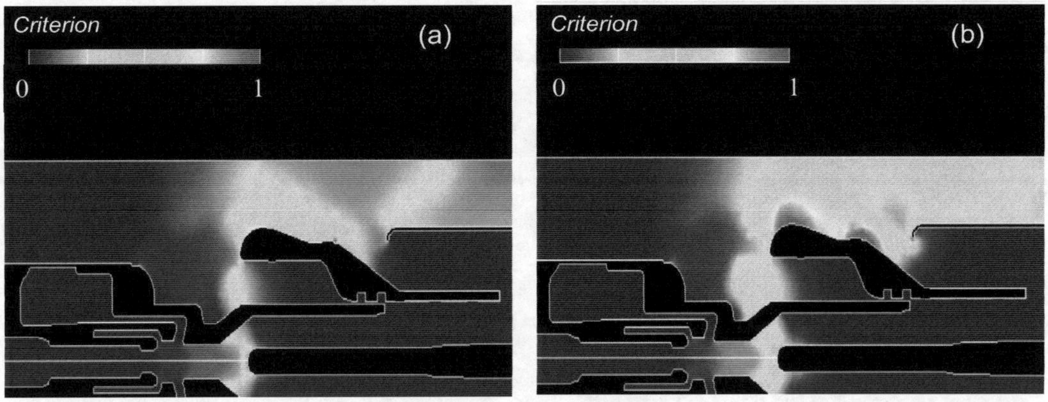

Figure 8: Withstand or breakdown criterion for 23 ms (a) and 25 ms (b) arcing times

The application refers to an air insulating gas switchgear tested for a terminal fault situation in SF_6 and for 2 arcing times: 23ms (successful test) and 25ms (test failure). The temperature field for 25ms is higher between the shield and the tank compared to the 23ms arcing time case. Consequently, the ratio between the electric field issued from the real transient recovery voltage and the theoretical dielectric withstand (or the dielectric breakdown field strength) of SF_6, calculated from Boltzmann equation solution, is given in figure 8. Comparison between these two results clearly shows that the 25ms configuration would lead to a failure, which has been observed during the tests. Moreover, the restrike point appears to be along the parallel shield as it has been observed during the test (fig 9). The compilation of such comparisons and new experimental measurements are being realised and will permit us in future works to define validity of such an approach.

Figure 9: Restrike traces along a shield

8. Conclusion:

Present paper is devoted the calculation of the critical electric field strength in the case of complex gas mixtures composing the hot SF_6. Such gas mixture can involve at least ten different molecules and atoms without any literature information on the dielectric strength (or the breakdown field) on the case of the most of them. Such information on dielectric strength can be obtained from the basic knowledge of the elastic and inelastic collision cross sections corresponding to the electron interaction with the different molecules and atoms of the complex hot gas mixture.

In the present paper, simplified formalisms are used to calculate the elastic and inelastic collision cross sections in the electron/gas system not available in the literature (as e.g. for e/SF_5 or e/SF_4, etc). The obtained sets of collision cross sections are then completed and validated using a swarm unfolding technique based on the comparison between calculated and measured swarm parameters. In the present paper, our electron swarm parameters are calculated using a multi-term Boltzmann equation solution. This is then followed by a second validation by comparing the sparse data on the electric field strength available in the case of a little number of SF_6 temperature and pressure.

Finally, this paper attempts to give an accurate idea on the usefulness of the electron-molecule cross sections in the case of the practical application of the dimensioning of the high voltage gas circuit breaker. This shows more particularly the interesting link between a fundamental research work on the electron interactions and a specific application to the gas circuit breaker behaviour during the dielectric recovery.

9. References:

[1] Ph. Teulet, 2004, Private communication.

[2] M Yousfi, Ph Robin Jouan, Z Kanzari, 2005, *IEEE Trans. on Dielectrics and Electrical Insulation,* 12, 6, 1192-1200

[3] B. Eliasson and E. Schade, 1985, "Electrical breakdown of SF_6 at high temperature (<2300K)", *Int. Conf. Phenomena in Ionized Gases*, Vol. 1, p. 409, 1977 ; E. Schade, "Recovery of switching srcs", *XVII Int. Conf. Phenomena in Ionized Gases Budapest*, 277.

[4] M. Nagata, I. Miyachi, Y. Yokoi and K. Isaka, 1980, "Breakdown characteristics of high temperature air and SF_6 gas", *Proc. 6th IEE Conf. Gas Discharges and Their Applications* -London : IEE, 78.

[5] L. Rothhardt, J. Mastovsky, J. Blaha, 1985, *J. Phys. D : Appl . Phys.*, Vol. 14, p. L205, 1981 ; L. Rothhardt, J. Mastovsky, J. Blaha, *J. Phys. D : Appl . Phys.*, 18, L15.

[6] T. Uchii, K. Iwata, H. Kawano, T. Nakamoto, and K. Suzuki, 2001, *IEEE Power-Engineering-Society-Winter-Meeting-Conference proceedings*, 1, 289.

[7] J. D. Yan, M.T.C. Fang and Q.S. Liu, 1997, *IEEE Trans. on Dielectric and Electrical Insulation,* 4, p. 114.

[8] G. Cliteur, Y. Hayashi, E. Haginomori and K. Suzuki, 1998, *IEEE Trans. on Dielectric Strength* 5, 843.

[9] L.G. Christophorou and J.K. Olthoff, 2000., *J. Phys. Chem. Ref. Data*, 29, 267-330

[10] L.G. Christophorou, 1984, *Electron molecule interactions and their applications*, Vol. 1, Academic Press, Orlando (84), US (book).

[11] T. M. Miller, A.E.S. Miller and J.F. Paulson, 1994, *J. Chem. Phys.*, 100, 8841-8848.

[12] C.H. Jackman, R.H. Garvey and A.E.S. Green, 1977, *J. Geophys. Res.*, 82, 5081-5090

[13]Y.-K. Kim,[1] K.K. Irikura,[2] M.E. Rudd,[3] M.A. Ali,[4] and P.M. Stone[1], J. Chang,[5] J.S. Coursey,[5] R.A. Dragoset,[5] A.R. Kishore,[5] K.J. Olsen,[5] A.M. Sansonetti,[5] G.G. Wiersma,[5] D.S. Zucker,[5] and M.A. Zucker, http://physics.nist.gov/PhysRefData/Ionization/intro.html

[14] M. Yousfi, S. Sigur and T. Vassiliadis, 1985, *J. Phys. D : Appl. Phys,* 18, 359-375.

[15] M. Hayashi and T. Nimura, 1983, *J. Appl. Phys,* 54, 4679-4882.

[16] S. Vacquie, A. Gleizes and M. Sabsabi, 1987 *Phys. Rev. A*, 35, 1615.

[17] T. Aschwanden, 1984 "Swarm parameters in SF_6 and SF_6/N_2 mixtures determined from a time resolved discharge study", *Gaseous Dielectrics IV*, edited by Christophorou (Pergamon, New York), 24-33

[18] Ph. Robin-Jouan, N. Kairouani, 2002, "Numerical and experimental analysis of the propagation of hot arc plasma in high voltage circuit-breakers", *IEE Conf on Gas Discharges and Their Applications* Liverpool, Vol. 1. 103-106.

Journal of Physics: Conference Series **115** (2008) 012013 doi:10.1088/1742-6596/115/1/012013

Computation of total electron scattering cross sections for molecules of astrophysical relevance

Minaxi Vinodkumar[1], Chetan Limbachiya[2], K N Joshipura[3] Bhushit Vaishnav[4] and Sumona Gangopadhyay[3]
[1]V. P. & R. P. T. P. Science College, Vallabh Vidyanagar-388 120, INDIA
[2] P. S. Science College, Kadi- 382 715, INDIA
[3]Department of physics, Sardar Patel University, Vallabh Vidyanagar-388 120, INDIA
[4]VPMP Polytechnic, Near Government ITI, Sector – 15, Gandhinagar – 382 015, INDIA

E-mail: minaxivinod@yahoo.co.in

Abstract. In this paper we report calculations of various total cross sections namely total elastic *(Q_{el})*, total ionization *(Q_{ion})*, total (complete) cross sections *(Q_T)* and grand total cross sections *(Q_{TOT})* for the impact of electrons on methyl compounds namely, methanol, methylamine and total ionization cross sections on reactive diatomic radical SO and SO_2 molecule at energies from circa threshold to 2000 eV. We have employed the 'Spherical Complex Optical Potential' (SCOP) formalism to evaluate $Q_{el,}$ $Q_{inel,}$ Q_T and Q_{TOT} and used our semi-empirical, 'Complex Spherical Potential – ionization contribution' (CSP-ic) method to derive Q_{ion}. These results are compared with available experimental and other theoretical data and overall good agreement is observed

1. Introduction

The list of astrophysically relevant molecules is long and growing. The study of dynamics and cross sections of electron molecule collisions forms the essential basis for understanding the characteristics of a target molecule. Moreover relations between electron scattering and physio-chemical properties of target molecules have been investigated since the earliest systematic measurements of these cross sections. These correlations indicate the role of target properties in the scattering process, while the semi-empirical formulae describing these correlations will help to estimate total cross sections of targets that are still lacking the experimental data or calculations.

The total cross sections including elastic plus all energetically possible inelastic channels for electron molecule systems are important in many applied sciences. Here we are interested in the intermediate and high-energy region (from ionization threshold up to 2 KeV) where almost all inelastic channels (rotational, vibrational and electronic excitation, ionization, dissociative processes etc) are open. Moreover, in order to develop understanding of the basic chemical behavior of above listed molecules of astrophysical relevance, the data on the total elastic, inelastic and ionization cross sections would prove crucial and therefore such study has attracted theoreticians and experimentalists in the last few decades.

Among the targets considered in this paper dipolar SO molecule is an example of a reactive diatomic radical. The SO radical is an abundant constituent of plasma containing sulfur dioxide. SO is readily produced by electron impact and ion impact dissociation and by photo dissociation of the parent SO_2 molecule [1-3]. Electron collisions with SO, in particular the electron-impact induced ionization and dissociation of SO, play an important role in a quantitative understanding of the fundamental collision processes and the plasma chemistry in SO_2 containing plasma. Electron collisions with the SO_2 molecule are important in many areas of application ranging from the SO_2 containing plasma torus around Jupiter and its satellite to diffuse-discharge switches to the plasma assisted treatment of biocompatible materials and biomedical surfaces [4].

Molecules involving carbon, hydrogen and oxygen have received much scientific interest in the field of molecular physics as prototypes of polyatomic molecules. Discovery of simple alcohols like methanol and methylamine in the interstellar space and in the atmosphere of planets in our own solar system has prompted the theoreticians and experimentalists to investigate electron impact cross sections in particular ionization cross sections. Further for In order to understand the behavior of these molecules in natural systems and in the laboratory, various reliable absolute cross sections are needed. Understanding of the formation of molecules containing carbon, nitrogen and hydrogen in Interstellar medium (ISM) is important because these are the building blocks of proteins and DNA [5].

In spite of the importance of these molecules, the information available for them is very scarce. Tarnovsky *et al* [1] have reported measurements of absolute partial ionization cross sections for the formation of parent and fragment ions resulting from the electron-impact ionization and dissociative ionization of SO free radical from threshold to 200 eV. Deustch *et al* [6] have calculated total single ionization cross sections using a modified additivity rule. Except these results there is no other data available in the literature for SO to the best of our knowledge. Hence the study of this molecule becomes important. Compared to SO molecule, SO_2 is studied more intensively. There are numerous experimental investigations compared to theoretical data [4]. The measurement of total cross section is reported by Szmytkowski and Maciag [7] while total ionization cross sections are reported by Basner *et al* [4] and Cadez *et al* [8].

Szmytkowski and Krzysztofowicz [9] have reported the total cross sections for methanol and methyl amine in the range 1 to 250 eV. Sueoka *et al* [10] have also reported the total cross sections for methanol. But no data either theoretical or experimental are available for total elastic cross sections for methanol. Duric *et al* [11] and Srivastava *et al* [12] have reported ionization cross sections for methanol. Very recently Rejoub *et al* [13] have determined the total and partial ionization cross sections for simple alcohols. For methyl amine the total cross sections are reported by Szmytkowski and Krzysztofowicz [9] and Schmieder [14]. Despite the importance of electron scattering from methylamine, not much is known about any theoretical calculations or experimental data for total elastic and total ionization cross sections. The total elastic cross sections for methanol is also not seen in the literature and hence probably reported here for the first time perhaps.

2. Theoretical Methodology

We briefly describe the theoretical formalism employed here to determine various total cross sections (TCS) for the impact of electrons on the targets studied. A more detailed discussion can be found in our earlier papers [15 -21]. Our aim in this paper is to calculate Q_{el}, Q_{ion} and Q_T

and to study their relative contributions to the total (complete) cross sections, Q_T for the chosen targets. Towards this goal, we have employed the well-known *'spherical complex optical potential'* (SCOP) formalism, through which the total elastic cross sections Q_{el} and its inelastic counterpart, Q_{inel} are obtained such that,

$$Q_T(E_i) = Q_{el}(E_i) + Q_{inel}(E_i) \tag{1}$$

For the polar molecules like CH_3OH and CH_3NH_2, we have also determined the dipole rotational excitation cross section $Q_{rot}(D, E_i)$. We can then define the grand total cross section by

$$Q_{TOT}(E_i) = Q_T(E_i) + Q_{rot}(D, E_i) \tag{2}$$

The dipole rotational excitation cross section $Q_{rot}(D, E_i)$ is calculated using the first Born approximation for a molecular dipole of strength *'D'* [15].

In the present range of electron energy (threshold to 2 *KeV*), many scattering channels that lead to discrete as well as continuum transitions in the target are open. Therefore we represent the electron-molecule system by a complex potential,

$$V(r, E_i) = V_R(r, E_i) + i\, V_I(r, E_i), \text{ such that} \tag{3}$$

$$V_R(r, E_i) = V_{st}(r) + V_{ex}(r, E_i) + V_{pol}(r, E_i) \tag{4}$$

The three terms on the RHS of the equation (4) represent the various real potentials arising from the electron target interaction namely, static, the exchange and the polarization potentials respectively. The most important basic input for evaluating all these potentials is the charge density of the target. This is obtained from the spherically averaged molecular charge-density $\rho(r)$, which is determined from the constituent atomic charge densities derived from the wave functions of Bunge and Barrientos [22]. The molecular $\rho(r)$ is renormalized to incorporate the covalent bonding as in our earlier papers [15-21].

For the exchange potential, we have used Hara's 'free electron gas exchange model' [23]. And for the polarization potential V_p, Zhang *et al* [24] has given a smooth form at all r, which has a correct asymptotic form ($-\alpha/2r^4$) at large *'r'* and approaches free electron-gas correlation energy $V_{co}(r)$ proposed by Perdew and Zunger [25] in the near target region as

$$V_P(r) = -\frac{\alpha}{2(r^2 + r_{CO}^2)^2} \tag{5}$$

where the constant r_{CO} can be determined by letting $V_P(0) = -\alpha/2r_{CO}^4 = V_{CO}(r=0)$ and α is the average dipole polarizability. This potential model has been proved fairly successful to TCS for electron atom scattering [25].

Finally, the imaginary part V_I of the complex potential is the absorption potential which represents approximately the combined effects of all inelastic channels. Here we have employed a well-known non-empirical quasi-free model form given by Staszeweska *et al* [26], see also Fujimoto and Lee [27]. Thus,

$$V_{abs}(r,E_i) = -\rho(r)\sqrt{\frac{T_{loc}}{2}}\cdot\left(\frac{8\pi}{10k_F^3 E_i}\right)\cdot\theta(p^2 - k_F^2 - 2\Delta)\cdot(A_1 + A_2 + A_3) \quad (6)$$

The local kinetic energy of the incident electron is

$$T_{loc} = E_i - (V_{st} + V_{ex}) \quad (7)$$

The absorption potential is not sensitive to long range potentials like V_{pol}. In equation (6), $p^2 = 2E_i$, $k_F = [3\pi^2 \rho(r)]^{1/3}$ is the Fermi wave vector and Δ is an energy parameter. Further $\theta(x)$ is the Heaviside unit step-function, such that $\theta(x) = 1$ for $x \geq 0$, and is zero otherwise. The dynamic functions A_1, A_2 and A_3 occurring in the equation (6) depend differently on $\rho(r)$, Δ, E_i and I (Ionization threshold of target). Detailed expressions of these functions are given in Staszeweska *et al* [26] and also in Fujimoto and Lee [27]. The energy parameter Δ determines a threshold below which $V_{abs} = 0$, and the ionization or excitation is prevented energetically. We have modified the original model, by considering Δ as a slowly varying function of E_i around I. A preliminary calculation is done with a fixed value $\Delta = I$, but Δ as a variable accounts for the screening of the absorption potential in the target charge-cloud region (Blanco & Garcia [28]), and also yields better agreement with experimental and other data in many cases. The modification introduced in our paper has been to assign a reasonable minimum value $0.8I$ to Δ and expressing this parameter as a function of E_i around I, as follows;

$$\Delta(E_i) = 0.8I + \beta(E_i - I) \quad (8)$$

Now let us say that E_p is the value of E_i at which our Q_{inel} attains maximum. In equation (8) β. is then obtained by requiring that $\Delta = I + 1$ (eV) at $E_i = E_p$, beyond which Δ is held constant. The expression for $\Delta\ (E_i)$, equation (8), is meaningful since Δ fixed at I would not allow even excitation at incident energy $E_i \leq I$. On the other hand, if parameter Δ is much less than the ionization threshold, then V_{abs} becomes unduly high near the peak position. In short the present form of $\Delta\ (E_i)$, equation (8) balances all these aspects and allows us to obtain satisfactory values of Q_{ion} for a given target. After generating the full complex potential given in Equation (3) for a given electron-target system, we solve the Schrödinger equation to get complex phase shifts which are used to find the cross sections given in Equation1.

The total inelastic cross section, Q_{inel} cannot be measured directly and this restricts the scope of the traditional calculations. The total inelastic cross section, Q_{inel} can not be measured directly, but the experimental quantity is total ionization cross section, Q_{ion}, which is contained in Q_{inel}. Q_{inel} can be partitioned into two main contributions viz.

$$Q_{inel}(E_i) = \sum Q_{exc}(E_i) + Q_{ion}(E_i) \quad (9)$$

where the first term is the sum over total excitation cross sections for all accessible electronic transitions. The second term is the total cross section of all allowed ionization processes induced by the incident electrons. The first term arises mainly from the low-lying dipole allowed transitions for which the cross section decreases rapidly at higher energies. Now, the second term in equation (9), electron impact ionization corresponds to infinitely many open channels, as against the electronic excitation, which comes from a small number of discrete scattering channels. The second term in equation (9), corresponds to transition to continuum state.

Therefore, ionization contribution in Q_{inel} increases as the incident energy increases, thereby making Q_{ion} to dominate in Q_{inel}. Thus from equation (9),

$$Q_{inel}(E_i) \geq Q_{ion}(E_i) \tag{10}$$

There is no rigorous way to project out Q_{ion} from Q_{inel}. But in order to achieve this, a reasonable approximation can be evolved by starting with a ratio function,

$$R(E_i) = \frac{Q_{ion}(E_i)}{Q_{inel}(E_i)} \tag{11}$$

Perhaps a first ever estimate of ionization in relation to excitation processes was made by Turner *et al* [29]. They concluded from semi-empirical calculations that in gaseous water (H_2O), ionization was more probable than excitation above ~30 eV. If σ_{ion} and σ_{exc} are the cross sections of ionization and excitation respectively, then above 100 eV [29],

$$\sigma_{ion} / (\sigma_{ion} + \sigma_{exc}) \approx 0.75 \tag{12}$$

Since the usual complex potential calculations do not help in determining ionization contribution from the inelastic cross section, we have introduced a method based on the equation (11). In *Complex Scattering Potential – ionization contribution* (CSP-*ic*) method, the energy dependence of $R(E_i)$ is given by the following relation [12-15]

$$R(E_i) = 1 - C_1 \left[\frac{C_2}{U + a} + \frac{\ln(U)}{U} \right] \tag{13}$$

where the incident energy is scaled to the ionization potential I through a dimensionless variable ,

$$U = \frac{E_i}{I} \tag{14}$$

Equation (13) involves dimensionless parameters C_1, C_2, and 'a', which are determined by imposing three conditions on the function $R(E_i)$ as discussed in earlier papers [15-21]. Specifically, we have $0 \leq R \lesssim 1$, such that

$$R(E_i) = 0 \text{ for } E_i \leq I \tag{15-a}$$

$$R(E_i) = R_p \text{ for } E_i = E_p \tag{15-b}$$

$$R(E_i) \cong 1 \text{ for } E_i > E_p \tag{15-c}$$

Here, $R_p \approx 0.7$ stands for the value of the ratio R at $E_i = E_p$. The choice of this value is approximate but physically justified. The peak position E_p (typically around 50 eV) occurs at an incident energy where the discrete excitation cross sections are on the wane, while the ionization cross section is rising fast, suggesting that the R_p value should be above 0.5 but still below 1.We follow the general trend observed in well known targets like Ne, Ar, O_2, N_2, CH_4 etc. Joshipura *et al* [18-19], Vinodkumar *et al* [15-17], Turner *et al* [29] and Karwasz *et al* [30], that near the peak of ionization, the contribution of the cross section Q_{ion} is about 70-80 % in the total inelastic cross section Q_{inel} and the same increases with energy. The choice of R_P follows from the general observation that at energies close to peak of ionization, the contribution of molecular Q_{ion} is about 70 – 80 % in the total inelastic cross sections Q_{inel}. For a number of stable molecules like O_2, H_2O, CH_4, SiH_4 etc for which the experimental cross sections Q_{ion} are known accurately, the ratio rises steadily as the energy increases above threshold. [17-19, 29, 30] This behavior is attributed to the faster fall of the first term $\sum Q_{exc}$ in equation (1). However the choice of R_p in equation (15) is not rigorous and it introduces uncertainty in the final results. For about two dozen atomic and molecular targets examined by us so far, the said uncertainty is found to be within the experimental errors ~10-15% generally involved in most of the ionization measurements.

3. Results and discussion

Here we have used the well known SCOP formalism to calculate the total elastic and total inelastic cross sections. They are added coherently to give us the total (complete) cross sections Q_T. Further, for the polar molecule, the rotational excitation cross section, Q_{rot}, is calculated separately using the first Born approximation and then added incoherently to the total cross section to evaluate the grand total cross section Q_{TOT}. Finally we applied our CSP-ic method to extract the total ionization cross sections from total inelastic cross sections. The summed total excitation cross sections are obtained as a byproduct vide Equation (9).

Total ionization cross sections, Q_{ion}, for e- SO, SO_2, CH_3OH and CH_3NH_2 and grand total cross section Q_{TOT} *(this is not in table 1!!)*, complete cross sections Q_T for CH_3OH and CH_3NH_2 are exhibited in Table 1 in **(Å^2)**. Also, the presently calculated Q_{TOT}, Q_T, Q_{el} and Q_{ion} for all the listed targets are plotted along with available other theoretical and experimental data. Low energy behavior of these cross sections is of much scientific interest because various elastic and inelastic phenomena occur around the threshold of the target. The sum total (complete) cross sections provide a useful sum-check of the different investigations. The CSP-*ic* method defined through the equations (8-15) has already been employed successfully to obtain total ionization cross sections of a number of atomic–molecular targets as demonstrated in Joshipura *et al* [19,21] and Vinodkumar *et al* [15-17].

Table 1: Various total cross sections for e-SO, SO_2, CH_3OH and CH_3NH_2 in (Å^2)

	Q_{ion}		CH_3OH				CH_3NH_2			
E_i (eV)	SO	SO_2	Q_{ion}	Q_{el}	Q_T	Q_{TOT}	Q_{ion}	Q_{el}	Q_T	Q_{TOT}
15	0.88	0.27	0.21	25.15	25.65	37.58	1.15	13.78	16.99	25.50
20	2.28	1.37	1.04	21.5	22.65	31.78	3.08	13.59	18.72	24.28
25	3.36	2.17	1.98	18.00	20.79	28.23	4.46	13.03	19.36	23.55
30	4.02	2.62	2.78	15.4	19.28	25.56	5.31	12.19	19.24	22.63
35	4.40	3.19	3.41	13.28	18.04	23.50	5.79	11.28	18.71	21.61
40	4.64	3.70	3.89	11.49	16.83	21.65	6.04	10.41	18.03	20.55
50	4.95	4.33	4.42	8.89	14.84	18.75	6.12	9.02	16.62	18.84
60	5.06	4.84	4.55	7.34	13.51	16.82	6.03	7.98	15.43	17.40
70	5.02	5.11	4.50	6.43	12.48	15.34	5.90	7.19	14.47	16.18
80	4.97	5.36	4.37	5.88	11.72	14.23	5.73	6.6	13.65	15.16
90	4.91	5.47	4.30	5.51	11.08	13.35	5.55	6.14	12.94	14.29
100	4.85	5.60	4.14	5.22	10.55	12.59	5.38	5.77	12.33	13.55

Journal of Physics: Conference Series **115** (2008) 012013 doi:10.1088/1742-6596/115/1/012013

200	4.15	5.22	3.04	3.86	7.51	8.58	4.13	4.01	9.17	9.81
300	3.60	4.63	2.43	3.23	6.02	6.74	3.43	3.31	7.58	8.02
400	3.18	4.11	2.05	2.82	5.10	5.65	2.95	2.88	6.73	7.07
500	2.86	3.68	1.77	2.52	4.46	4.91	2.60	2.58	5.87	6.09
600	2.60	3.33	1.56	2.29	3.98	4.35	2.33	2.35	5.15	5.38
700	2.39	3.04	1.41	2.09	3.59	3.92	2.11	2.16	4.65	4.84
800	2.21	2.79	1.27	1.94	3.29	3.29	1.93	2.01	4.24	4.41
900	2.06	2.58	1.16	1.81	3.04	3.04	1.78	1.88	3.91	4.06
1000	1.93	2.40	1.09	1.69	2.82	2.82	1.65	1.77	3.63	3.77
2000	1.16	1.36	0.64	0.97	1.63	1.63	0.97	1.15	2.40	2.47

Consider first the exotic free SO radical. In Fig.1, we have plotted total ionization cross sections for SO and compared with the available results of Tarnovsky *et al* [1]. Very good agreement is observed between the measured values of Tarnovsky *et al* and present results from threshold to 100 *eV*, but above 100 *eV* the values of Tarnovsky *et al* [1] are lower than our values. The calculated total single ionization cross sections [1] using the modified additivity rule for (SO^+) are shown by the dashed curve in Figure 1.

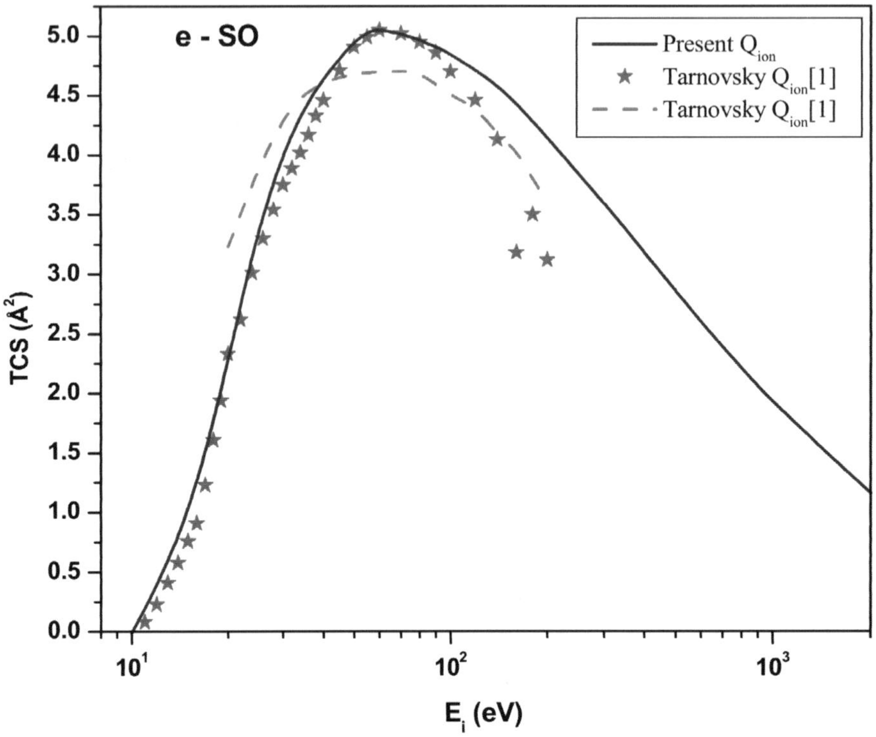

Fig 1 : Total ionization cross sections for e – SO scattering

Fig. 2 shows the comparison of total ionization cross sections of SO_2 molecule on electron impact. From threshold to 50 eV there is very good agreement of present results with theoretical values of Kim *et al* [31] and the experimental values of Cadez *et al* [8]and Basner *et al* [4]. But above 50 eV, present results are in good accord with experimental values of Cadez *et al* but they are lower compared to theoretical values of Kim *et al* [30] and experimental values of Basner *et al* [4]. Basner *et al* have reported 18 – 25% error in their measurements and our results fall within their experimental uncertaintites.

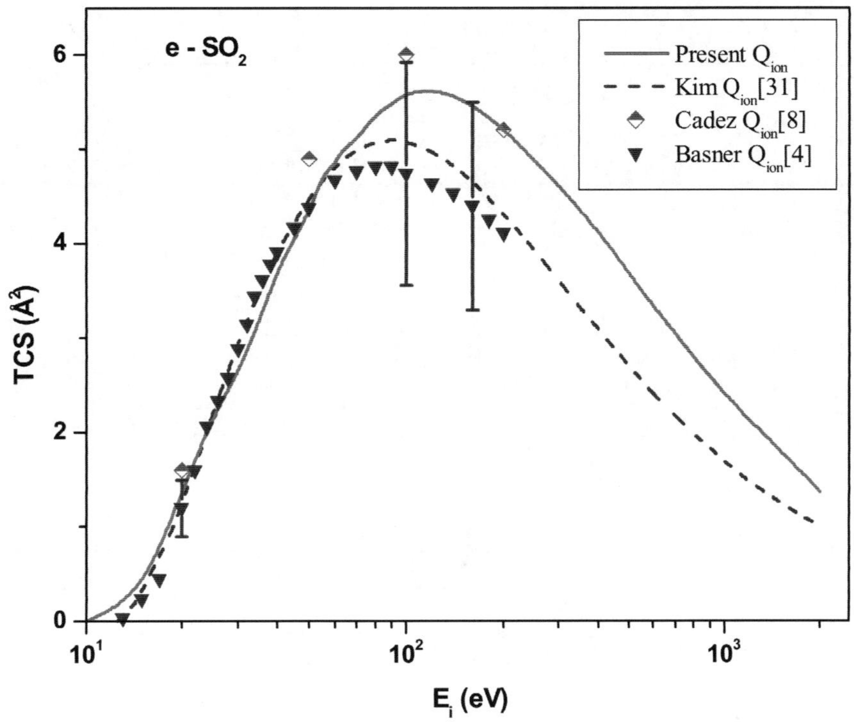

Fig 2 : Total ionization cross sections for e – SO_2 scattering

Consider next methanol which is more widely studied molecule compared to other targets investigated here. Fig 3 shows the comparison of grand total cross section Q_{TOT},total (complete) cross sections Q_T and total ionization cross sections Q_{ion} for e- CH_3OH scattering. The present Q_{TOT} is found to be in good agreement with the experimental values of Szmytkowski and Krzysztofowicz [9] and Sueoka *et al* [10] between the energies 50 -200 eV while at lower energies the experimental values are quite lower than our Q_{TOT} values. This discrepancy especially at lower energies can be attributed as mentioned by Itikawa and Mason [32] in connection with *e-* H_2O scattering, to the dipolar nature of the target that affects measurements of complete total cross sections Q_{TOT}. The present total ionization cross sections are in excellent agreement with available experimental data [11] upto 100 *eV* beyond which their values are slightly higher than our values. There is no data either theoretical or experimental for total elastic cross sections in the literature to the best of our knowledge.

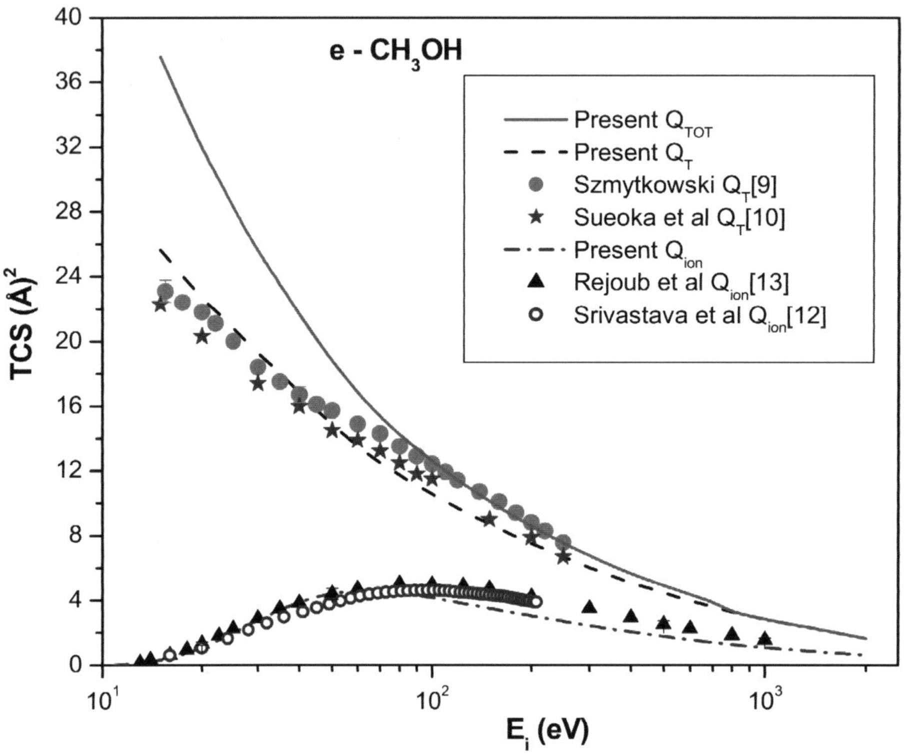

Fig 3 : Various total cross sections for e – CH_3OH scattering

Finally Fig 4 shows the dependence of various total cross sections as function of incident energy for *e – CH_3NH_2* scattering. CH_3NH_2 is a polar molecule with permanent dipole moment of 1.31 Deybe '*D*' [6]. Hence it gives rise to rotational excitation and the rotational cross sections are significant at lower energies below 100 *eV* which can be noted in the figure. At low energies there is clear distinction between Q_T and Q_{TOT} curve while at high energies both curves match reflecting the fact that at very high energy there is negligible contribution to the total cross section from the rotational excitation. Despite the importance of methylamine there is not much work found in the literature. However for $Q_{T,}$ there are experimental results of Szmytkowski and Krzysztofowicz [6] from low energy to 250 *eV*. There is good agreement between present Q_{TOT} and experimental values of Szmytkowski and Krzysztofowicz between 50 *eV* to 250 *eV*. Below 50 *eV* there is a discrepancy between the two results. There is no comparison either theoretical or experimental for total elastic as well as total ionization cross sections to best of our knowledge. There is probably no theoretical or experimental data for total elastic and total ionization cross sections found in the literature.

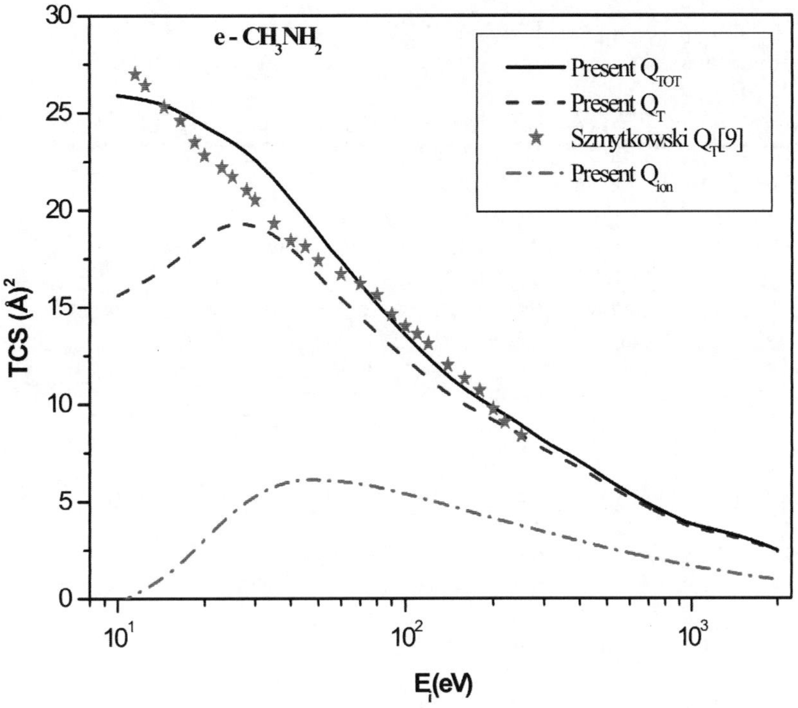

Fig 4 : Various total cross sections for e – CH_3NH_2 scattering

4. Conclusion

In conclusion, this paper presents our calculations on electron impact cross sections Q_{TOT}, Q_T, Q_{el}, Q_{inel} and Q_{ion} for the methyl compounds and total ionization cross sections Q_{ion} for diatomic radical SO and SO_2 molecule. The resulting Q_{TOT} values are reliable and agree in general. The CSP-*ic* method adopted here though not rigorous is capable of yielding reasonable total ionization cross section as a fraction of total *inelastic* cross section at a given incident energy. The present comparison can help in obtaining a recommended data set for listed molecules. Besides this, we can also conclude about the relative importance of various collision processes like elastic scattering, ionization etc, induced by electrons. As for the lesser known SO radical and CH_3NH_2 molecule, it is hoped that the present work will stimulate experimental and further theoretical research on electron scattering and ionization.

Acknowledgement

MVK and CGL are thankful to University Grants Commission New Delhi for the major research project under which part of this work is carried out. A part of this research was funded through a Project awarded to KNJ by Indian Space Research Organization, (Banglore) India.

References:

[1] Tarnovsky V, Levin A, Deutsch H and Becker K 1995, *J. Chem. Phys.* **102(2)** 770.
[2] Gallinberti L 1988, *Pure Appl. Chem.* **60** 663
[3] Chang M B, Kushner M J and Rood M J 1992, *Plasma Chem. Plasma Proc.* **12** 565.
[4] Basner R, Schmidt M, Deutsch H, Tarnovsky V, Levin A and Becker K 1995, *J. Chem. Phys.*, **103(1)** 211
[5] Pele A, Sailer W, Scheir P, Mason N J , Illenberger E and Mark T D 2003, *Vacuum* **70** 429
[6] Deustch H, Märk T. D, Tarnovsky V, Becker K, Comelissen C, Cespiva L, Bonacic-Koutecky V, *Int. J. Mass Spectrom. Ion. Proc.*,**137** (1994) 77.
[7] Szmytkowski Cz and Maciag K 1986, *Chem. Phys. Lett.* **129** 321
[8] Cadez I M, Pejcev V M, Kurepa M V 1983, *J. Phys. D: Appl. Phys.* **16** 305
[9] Szmytkowski Cz and Krzysztofowicz A M 1995, *J. Phys. B: At. Mol. Opt. Phys.*, **28** 4291
[10] Sueoka O, Katayama Y and Mori S 1985, *At. Colli. Res. Japan, Prog. Rep.* **11** 17
[11] Duric N, Cadez L and Kurepa M 1989, *Fizika (Zagreb)* **21** 319.
[12] Srivastava S K, Krishnakumar E and Fucaloro A F and Note T. Van 1996, *J. Geophys.* Res. **101** 26155
[13] Rejoub R, Morton C D, Lindsay B G and Stebbings R F 2003, *J. Chem. Phys.* **118(4)** 1756
[14] Schmieder F 1930, *Elektrochem.* **36** 700
[15] Vinodkumar M, Limbachiya C, Antony B and Joshipura K N 2007, *J. Phys. B: At. Mol. Opt. Phys.* **40** 3259
[16] Vinodkumar M, Joshipura K N, Limbachiya C and Antony B K 2006, *Eur. Phys. J.* D **37** 67
[17] Vinodkumar M, Joshipura K N, Limbachiya C, and Mason N 2006, *Phys. Rev.* A, 74 022721
[18] Joshipura K N, Vinodkumar M, Limbachiya C G and Antony B K 2004, *Phys. Rev.* A **69** 022705
[19] Joshipura K N, Vinodkumar M, Antony B K, and Mason N J 2003, *Eur. Phys. J.* D **23** 81
[20] Vinodkumar M, Joshipura K N, Limbachiya C G, and Antony B K 2003, *Nucl. Instr. and Meth. In Phys. Res.* B **212** 63
[21] Joshipura K N, Antony B K, and Vinodkumar M 2002, *J. Phys. B: At. Mol. Opt. phys.* **35** 4211
[22] Bunge C F and Barrientos J A 1993 *At Data Nucl. Data Tables* **53** 113
[23] Hara S 1967, *J. Phys. Soc. Japan* **22** 710
[24] Zhang X , Sun J , Liu Y 1992, *J. Phys. B: At. Mol. Opt. Phys.* **25** 1893
[25] Perdew J P, Zunger A 1981, *Phys. Rev.* B **23** 5048
[26] Staszewska G, Schwenke D W, Thirumalai D and Truhlar D G 1983 *Phys. Rev.* A **28** 2740
[27] Fujimoto M M and Lee M T 2000 *J. Phys. B: At. Mol. Phys.* **33** 4759
[28] Blanco F and Garcia G 2003 *Phys. Rev.* A **67** 0022701
[29] Turner J E, Paretzke H G, Hamm R N, Wright H A and Richie R H 1982 *Radiat. Res,.USA* **92** 47
[30] Karwasz G P, Brusa R S and Zecca A 2001 *La Rivista Del Nuovo Cimento* **24** 1
[31] Kim Y-K, Hwang W, Weinberger N M, Ali M A and Rudd M E 1997, *J. Chem.Phys.* **106** 1026
[32] Itikawa Y and Mason N J, *J. Phys. Chem. Ref. Data*, **34** (2005) 1

Journal of Physics: Conference Series **115** (2008) 012014 doi:10.1088/1742-6596/115/1/012014

Simulations of the fragmentation of the [V-H]$^-$ anions as formed upon DEA to L-valine.

Flosadóttir HD, Jónsson H, Ingólfsson O
Science Institute, University of Iceland, Dunhagi 3, 107 Reykjavík, Iceland

odduring@hi.is

Abstract. The dissociation of proteins due to radiation and charged particles is an important process which is not well understood. Measurements of the fragmentation of such large molecules using mass spectrometers can be difficult to describe and understand since for many of the masses observed there are more than one possible combination of atoms. We have simulated the metastable decay of the anions formed by dissociative electron attachment (DEA) to L-Valine to show that it is possible to recognize some of the observed mass to charge ratios by their further fragmentation. By using computer simulations of the classical dynamics of the ions and finding minimum energy paths for intermolecular atom transfer and reorganization, we have identified molecular combinations that can be assigned to the measured m/Z ratios. The atomic forces in these calculations are obtained from density functional theory using a gradient dependent functional.

1. Introduction

The damage to biologically relevant molecules due to high energy radiation is recognized to be the result of a complex process of sequential reactions. Though electro-magnetic radiation of sufficient energy can cause direct bond rupture the biological damage caused by high energy radiation traversing living organism is mainly caused by secondary species formed along the radiation track. These species may be ions, radicals or free electrons. It has long been accepted as a fact, that the radical species formed play the dominating role in initiating the chemistry that causes the actual damage. However, in recent years the attention has turned increasingly towards the role of low energy electrons in this process. These low energy electrons (<20 eV) are the most abundant species along the track [1]. Typically the yield is about 5×10^4 per MeV of the incident radiation [2].

Low energy electron attachment is a process that has been studied on a variety of molecular species throughout the last decades. The first step is the formation of a transient negative ion (TNI).

$$E^- + ABC \rightarrow ABC^{-\#}. \quad (1)$$

This TNI can then dissociate directly along the corresponding dissociative asymptote of the anion to form a negatively charged fragment and its neutral radical counterpart. The initial formation of the TNI may be described within the frame of the Born-Oppenheimer approximation as a vertical transition from the electronic ground state of the neutral molecule to the potential energy surface of the anion formed (Franck-Condon transition).

The dissociation of the anion formed can often be described within the framework of a diatomic dissociation. This is especially true when the molecule in question is fairly simple and the incident energy of the electron low; i.e.,

$$ABC^{-\#} \rightarrow AB^{(\#)} + C^{-}. \quad (2)$$

However, as the molecule becomes more complex more and more processes are observed that may only be described by considerable rearrange within the molecule.

$$ABC^{-\#} \rightarrow AC^{(\#)} + B^{-}. \quad (3)$$

Such pre-dissociation processes are commonly associated with an initial occupation of a π^* orbital as has for example been observed for the CN^- formation from the aliphatic amino acids at low electron energies [3-6]. This is a process where four bonds are broken and at least two new bonds must be formed to account for the energy needed.

Substantial work has been done on dissociative electron attachment to the individual building blocks of essential biological molecules such as the DNA, the RNA and proteins and peptides. This includes the nucleobases[7-12], ribose[13,14] and deoxyribose[15], phosphodiesters[16] and a number of amino acids[3-6,17-19]. These experiments show clearly that low energy electrons can cause substantial damaged to biological molecules. This has also been clearly demonstrated by low energy electron irradiation of plasmid DNA condensed on a surface. These experiments show that low energy electrons can cause both single- and double-strand breaks through a resonant attachment process[20].

Though it is clear in the meanwhile that low energy electrons can cause substantial damage to biologically relevant molecules it is less established how the damaging bond ruptures proceed. Even the assignment of m/Z ratios observed upon DEA to the smaller building blocks is often not unambiguous and high resolution mass spectrometry or isotope labeled molecules must be used to establish the actual reaction path[8,9,21-26]. Furthermore, it is often not trivial with the currently used experimental setups to distinguish between direct dissociation upon electron attachment and metastable decay processes, i.e., the further fragmentation of a negative ion fragment formed upon DEA. In the case of complex bio-molecules the large fragments initially formed through DEA can accommodate a substantial amount of internal energy in their vibrational degrees of freedom and further fragmentation of such pre-dissociated ions may take place a comparatively long time after the initial dissociation.

Here we present dynamic calculations on the further fragmentation of the dehydrogenated valine anion $[V\text{-}H]^-$ as formed by DEA. Dissociative electron attachment to L-valine was recently reported by Papp et al. showing that $[V\text{-}H]^-$ is the most abundant anionic fragment formed. This is also the case for the other aliphatic amino acids[3-6,19]. In addition to $[V\text{-}H]^-$ Papp et al. observed the formation of 6 other fragments which can partly not be unambiguously assigned to one molecular composition[6].

Dynamical simulations were performed on the $[V\text{-}H]^-$ system for three possible hydrogen abstraction sites; the hydroxyl group, the α-C and the amino group, here after denoted by subscript as $[V\text{-}H]_O^-$, $[V\text{-}H]_{\alpha C}^-$ and $[V\text{-}H]_N^-$, respectively. From our calculations we can see that the $[V\text{-}H]^-$ anionic fragments formed by DEA to the amino acid L-Valine at about 5 eV electron incident energy are metastable when the hydrogen extraction takes place from the amino group. In good agreement with previous experiments our calculations predict that the $[V\text{-}H]_N^-$ anion fragments further to form the anionic fragment $COOH^-$ and the neutral counterpart $(CH_3)_2CH\text{-}CH(NH_2)$- if the hydrogen abstraction takes place from the amino group but is stable with respect to further decay if the hydrogen abstraction takes place from the hydroxyl group or the α-C. In addition to the fragmentation channel leading to the $COOH^-$ formation our dynamic calculations also reveal a stabilization mechanisms through hydrogen transfer from the hydroxyl to the amino group. The minimum energy reaction path for the hydrogen transfer is calculated using the elastic band method [27].

2. Methodology

The basic assumption in the calculations performed here is that the anion formed after electron attachment is long lived enough to reach the electronic ground state and that the excess energy brought in by the electron gets distributed evenly over the vibrational degrees of freedom of the anion. The calculations are used to explore what secondary fragmentation can occur if this state is reached.

We use density functional theory (DFT) with the generalized gradient approximation functional PW91 in combination with a plane wave basis set and ultrasoft pseudopotentials[28] for the calculations of the electronic degrees of freedom. The plane wave basis set includes kinetic energy up to 400 eV. The VASP code was used for these calculations [29].

The nuclear degrees of freedom were treated in two ways, first by carrying out classical dynamics, and then by finding minimum energy paths for transitions. In both cases the atomic forces were obtained from DFT calculations. The dynamical simulations were done in six steps:

1) The geometries of the charged fragments were optimized starting from an initial guess by minimizing the energy.
2) Internal energy of 1 eV was added, equally distributed over the vibrational degrees of freedom, to account for the temperature used in the experiments to sublime the valine powder (400 K).
3) The heated system is allowed to equilibrate in a canonical simulation for T=400 K over a period of 100 fs, giving it time for about 10 vibrations.
4) Additional 5 eV are added to the system to account for the kinetic energy of the attaching electron.
5) A constant energy, micro canonical ensemble is simulated over a time period of 500 fs.
6) At last, the charge density is analysed using Bader's method [30,31].

For each fragment 10 starting points were taken from step 3). By this procedure we have reached the state the anions are in some while after initial dissociation, already having distributed the internal energy it acquired and equilibrated itself via vibrations.

For the calculations of the minimum energy path of the hydrogen transfer we used the nudged elastic band method[27]. There, 19 images of the system were equally distributed between the minima on each side of the transfer, giving a discrete representation of the path. The initial state is $[V\text{-}H]_N^-$ and at the final state $[V\text{-}H]_O^-$ where the subscript indicates the site lacking the proton.

By using plane wave basis sets in the DFT calculations and only including valence electrons, the dynamics simulations and the minimum energy path calculations are quite straight forward and not so computationally demanding. They are feasible for even larger systems and increased fragmentation and fluctuations.

3. Results and discussion

The most significant m/Z ratio formed by direct dissociation of valine upon DEA is 116 amu, i.e., the $[V\text{-}H]^-$ anion formed by hydrogen loss from the initially formed transient negative ion. This ion is formed through two distinct resonances; one at 1.2 eV and another one at 5.3 eV[6]. The most probable sites for the hydrogen loss from the transient negative ion are the hydroxyl-group, the α-C and the amino-group. The theoretically estimated value at the B3LYP 6-311+G** level for the energy threshold for hydrogen abstraction from the hydroxyl group and the α-C is 1.04 eV and 2.20 eV, respectively[6]. At the same level of theory we calculate the threshold for the hydrogen abstraction from the amino group to be 2.85 eV. Hence, the O-H bond energy is significantly lower than the other two and at 1.2 eV electron energy this is the only hydrogen abstraction channel that is energetically available.

This is also reflected in recent experiments on metastable decay of deprotonated valine anions in MALDI as compared to the metastable decay of the dehydrogenated valine anion up on DEA and collision induced dissociation measurements. These experiments show that the two resonances are not the same anion. In these experiments, the low energy resonance at 1.2 eV does not show any secondary fragmentation while the 5.3 eV resonance dissociates further by metastable decay to form the m/Z ratios 45 amu and 70 amu[26]. By considering the total energy, the lower resonance can only be due to loss of hydrogen from the hydroxyl group, but the second one is energetic enough for all of the hydrogen abstraction sites to be possible.

Here, we use DFT calculations to study from which site the hydrogen abstraction through the high energy resonance at 5.3 eV takes place. The dynamical simulations performed on the system were done on hydrogen loss from all three sites, i.e., starting with the $[V\text{-}H]_O^-$, $[V\text{-}H]_N^-$ and $[V\text{-}H]_{\alpha C}^-$ anions. Of the ten simulations done on each of those DEA fragments, none of the $[V\text{-}H]_O^-$ nor $[V\text{-}H]_{\alpha C}^-$ show any further fragmentation. $[V\text{-}H]_N^-$ on the other hand fragments into $COOH^-$ and [V-COOH] in 7 cases out of ten and the other three show a hydrogen transfer from O to N.

Figure 1 shows four snapshots from the simulations on $[V\text{-}H]_N^-$ (a) where further fragmentation of $[V\text{-}H]_N^-$ leads to the formation of $COOH^-$ and the neutral counterpart [V-COOH], and (b) where stabilization is achieved through hydrogen transfer from the hydroxyl group to the amino group to form the stable $[V\text{-}H]_O^-$ anion.

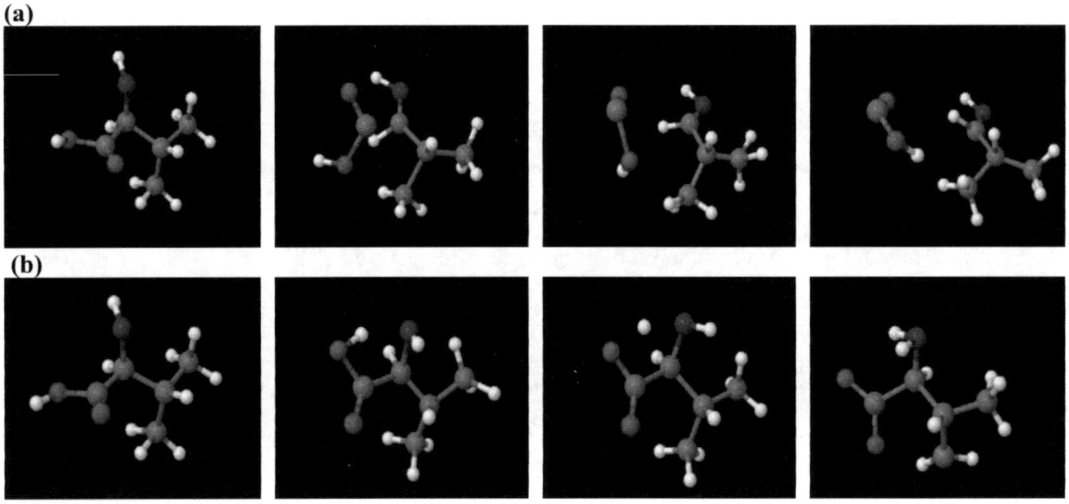

Figure 1. Simulated dynamics of $[V\text{-}H]_N^-$. (a) Formation of $COOH^-$ (m/Z = 45 amu) and the complementary neutral fragment [V-COOH]. (b) Proton transfer from the hydroxyl group to the amino group. The $[V\text{-}H]_O^-$ ion is stable and did not dissociate during the time interval simulated.

As is demonstrated in Fig. 1, our dynamic calculations are capable of reproducing the dissociation as observed from metastable decay of $[V\text{-}H]_N^-$ formed by hydrogen abstraction in DEA as well as through deprotonation in MALDI. In both cases the fragment m/Z = 45 amu is one of the two dominating fragments, which is in good agreement with our observation of the $COOH^-$ formation. The predicting power of the dynamic calculations is also supported by the absence of any further fragmentation of $[V\text{-}H]_O^-$ and $[V\text{-}H]_{\alpha C}^-$ as is to be expected from the experimental results.

Furthermore, our dynamic simulations also reveal channels that can not be experimentally observed, i.e., the hydrogen transfer from the hydroxyl group to the dehydrogenated amino-group. This quite interesting channel shows that the ion created is not necessarily the ion observed even though the m/Z ratio is unchanged. In good agreement with our observations for the precursor $[V\text{-}H]_O^-$ and $[V\text{-}H]_N^-$, this structural change makes the ion more stable and no further fragmentation is observed.

The rate of such a process is largely governed by the energy barriers along the minimum energy path. Assuming that the vibrations of the ion have thermally equilibrated and that the harmonic approximation is accurate enough, the activation energy (AE) is given by the highest energy along the minimum energy path. Figure 2 shows the minimum energy reaction path for the hydrogen transfer calculated by using the nudged elastic band method [27].

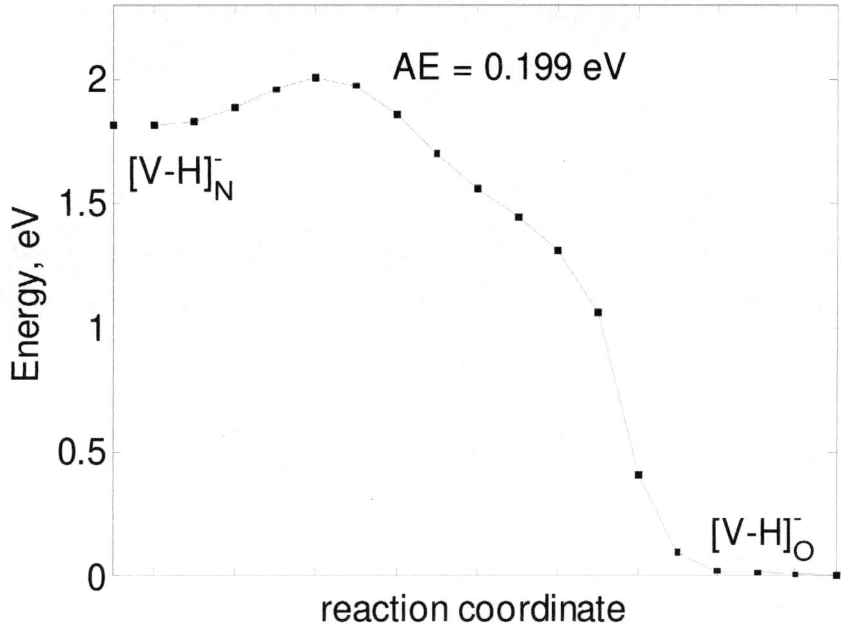

Figure 2. The energy along the minimum energy path for the hydrogen transfer. To the left is the $[V\text{-}H]_N^-$ (1.809 eV) and to the right is the $[V\text{-}H]_O^-$ (set as 0.000 eV). The data points show the energy of intermediate configurations used in the calculation. They are connected by a line for clarity.

The minimum energy path for the hydrogen atom transfer is calculated from the $[V\text{-}H]_N^-$ initial state to the $[V\text{-}H]_O^-$ final state, which is 1.809 eV lower in energy. The path (shown in Fig 2) was represented by 19 intermediate configurations and the results of the optimization are shown in Fig. 2. The maximum energy along the reaction coordinate was found to be at 2.008 eV compared to the energy of $[V\text{-}H]_O^-$. This results in an activation energy of 0.199 eV. Such a low energy barrier will be overcome quite readily given the large vibrational energy in the molecule. Figure 2 shows that the minimum energy path has a shoulder to the right of the maximum. This shoulder is because of a stabilization of the anion that occurs when the oxygen and the nitrogen share the hydrogen and a strained hydrogen bond is formed.

4. Conclusion

From these calculations we can see that some of the anionic fragments of the amino acid L-Valine are metastable, and have their specific and characteristic secondary fragmentation products. In good agreement with experiment we find that from the fragments $[V\text{-}H]_O^-$, $[V\text{-}H]_{\alpha C}^-$ and $[V\text{-}H]_N^-$ which all have the same m/Z ratio only $[V\text{-}H]_N^-$ fragments further. With our simulations we observe the characteristic decay of this ion, suggesting that our theoretical approach has the potential of becoming a valuable tool to aid the interpretation of DEA data by helping to identify the primary fragments through their characteristic secondary fragmentation. Furthermore, we also observe stabilization of the $[V\text{-}H]_N^-$ through hydrogen transfer from the hydroxyl group showing the potential this approach has as a tool to identify reaction channels that can not be observed with current experimental setups.

[1] International Comission on Radiation Units and Measurements, Washington, DC, 1979.
[2] Cobut V, Frongillo Y, Patau J P, Goulet T, Fraser M J, Jay-Gerin J P, 1998 *Radiat. Phys. Chem.* **51** 229-43.
[3] Ptasińska S, Denifl S, Candori P, Matejčík S, Scheier P, Märk T D, 2005 *Chem. Phys. Lett.* **403** 107-12.
[4] Gohlke S, Rosa A, Illenberger E, Bruning F, Huels M A, 2002 *J. Chem. Phys.* **116** 10164-9.
[5] Vasil'ev Y V, Figard B J, Voinov V G, Barofsky D F, Deinzer M L, 2006 *J. Am. Chem. Soc.* **128** 5506-15.
[6] Papp P, Urban J, Matejčík S, Stano M, Ingolfsson O, 2006 *J. Chem. Phys.* **125**.
[7] Hanel G, Gstir B, Denifl S, Scheier P, Probst M, Farizon B, Farizon M, Illenberger E, Märk T D, 2003 *Phys. Rev. Lett.* **90** 188104.
[8] Ptasińska S, Denifl S, Scheier P, Illenberger E, Märk T D, 2005 *Angew Chem Int. Ed.* **44** 6941-3.
[9] Ptasińska S, Denifl S, Mroz B, Probst M, Grill V, Illenberger E, Scheier P, Märk T D, 2005 *J. Chem. Phys.* **123**, 124302.
[10] Abdoul-Carime H, Langer J, Huels M A, Illenberger E, 2005 *Eur. Phys. J. D* **35** 399-404.
[11] Abouaf R, Dunet H, 2005 *Eur. Phys. J. D* **35** 405-10.
[12] Scheer A M, Aflatooni K, Gallup G A, Burrow P D, 2004 *Phys. Rev. Lett.* **92**, 068102.
[13] Baccarelli I, Gianturco F A, Grandi A, Sanna N, Lucchese R R, Bald I, Kopyra J, Illenberger E, 2007 *J. Am. Chem. Soc.* **129** 6269-77.
[14] Bald I, Kopyra J, Illenberger E, 2006 *Angew. Chem. Int. Ed.* **45** 4851-5.
[15] Ptasińska S, Denifl S, Scheier P, Märk T D, 2004 *J. Chem. Phys.* **120** 8505-11.
[16] König C, Kopyra J, Bald I, Illenberger E, 2006 *Phys. Rev. Lett.* **97**, 018105.
[17] Aflatooni K, Hitt B, Gallup G A, Burrow F D, 2001 *J. Chem. Phys.* **115** 6489-94.
[18] Abdoul-Carime H, Gohlke S, Illenberger E, 2004 *Phys. Chem. Chem. Phys.* **6** 161-4.
[19] Ptasińska S, Denifl S, Abedi A, Scheier P, Märk T D, 2003 *Anal. Bioanal. Chem.* **377** 1115-9.
[20] Boudaiffa B, Cloutier P, Hunting D, Huels M A, Sanche L, 2000 *Science* **287** 1658-60.
[21] Abdoul-Carime H, Gohlke S, Illenberger E, 2004 *Physical Review Letters* **92,** 168103.
[22] Ptasińska S, Denifl S, Grill V, Märk T D, Illenberger E, Scheier P, 2005 *Phys. Rev. Lett.* **95,** 093201.
[23] Sulzer P, Ptasińska S, Zappa F, Mielewska B, Milosavljević A R, Scheier P, Märk T D, Bald I, Gohlke S, Huels M A, Illenberger E, 2006 *J. Chem. Phys.* **125**, 044304.
[24] Ptasińska S, Denifl S, Grill V, Märk T D, Scheier P, Gohlke S, Huels M A, Illenberger E, 2005 *Angew. Chem. Int. Ed.* **44** 1647-50.
[25] Mauracher A D S, Aleem A, Wendt N, Zappa F, Cicman P, Probst M, Märk TD, Scheier P, Flosadóttir HD, Ingólfsson O, Illenberger E, 2007 *Phys. chem. chem. phys.* **9**(42) 5680
[26] HD Flosadóttir S D, F Zappa, N Wendt, A Mauracher, A Bacher, H Jónsson, TD. Märk, Pl Scheier, O Ingólfsson, 2007 *Angew. Chem. Int. Ed.* **46** (2007) 8057.
[27] Henkelman G, Jonsson H, 2000 *J. Chem. Phys.* **113** 9978-85.
[28] Vanderbilt D, 1990 *Physical Review B* **41** 7892-5.
[29] Kresse G, Furthmuller J, 1996 *Phys. Rev. B* **54** 11169-86.
[30] Henkelman G, Arnaldsson A, Jónsson H, 2006 *Comp. Mater. Sci.* **36** 354-60.
[31] Sanville E, Kenny S D, Smith R, Henkelman G, 2007 *J. Comp. Chem.* **28** 899-908.

Journal of Physics: Conference Series **115** (2008) 012015 doi:10.1088/1742-6596/115/1/012015

Electron-induced chemistry of alcohols

Bogdan C. Ibanescu, Olivier May, Michael Allan

Department of Chemistry, University of Fribourg, Chemin du Museé 9, 1700 Fribourg, Switzerland

E-mail: bogdancatalin.ibanescu@unifr.ch

Abstract. We studied dissociative electron attachment (DEA) to a series of compounds with one or two hydroxyl groups. For the monoalcohols we found, apart from the known fragmentations in the 6 − 12 eV range proceeding *via* Feshbach resonances, also new weaker bands at lower energies, around 3 eV. They have a steep onset at the dissociation threshold and we assigned them as proceeding *via* shape resonances with temporary occupation of $\sigma^*_{\text{O-H}}$ orbitals. These low energy fragmentations become much stronger in the larger molecules and even the strongest DEA process in the compounds with two hydroxyl groups. Above 6 eV, in the Feshbach resonance regime, the dominant process is a fast loss of a hydrogen atom from the hydroxyl group. In some cases the resulting $(M-1)^-$ anion (loss of hydrogen atom) is sufficiently energy-rich to further dissociate by loss of stable, closed shell molecules like H_2 or ethene.

1. Introduction

The interest in electron-induced chemical processes has been renewed by the discovery that electrons at subionization [1] and even subexcitation energies [2] damage DNA and by the emerging applications in materials science [3]. These applications inspired studies of the corresponding elementary step, the DEA to isolated polyatomic molecules in the gas phase and in particular to molecules of biological interest, with many interesting findings [4–7]. In our earlier work we studied several saturated compounds containing the amino and the hydroxy groups which could be considered as model compounds for biomolecules [8, 9]. Saturated compounds do not have π^* shape resonances and fragmentations are generally mediated by Feshbach resonances in the 6 − 12 eV region. Feshbach resonances with occupation of Rydberg-like orbitals are well known from scattering experiments other than DEA, in particular ETS of rare gases, diatomic and polyatomic molecules [10, 11] or in the vibrational excitation cross sections [12]. The energies of these Feshbach resonances were found to have a simple relation to the energies of the parent Rydberg states and the grandparent state of the cation[13]. The relation depends only weakly on the molecule in question and can be used to identify and assign the Feshbach resonances. This paper presents dissociative electron attachment spectra of a series of alcohols and assigns the observed bands to various types of resonances, guided by comparison with HeI photoelectron spectra and vibrational excitation cross sections.

2. Methods

The dissociative electron attachment spectrometer used to measure the yield of mass-selected stable anions as a function of electron energy has been described previously [14, 15]. The

Journal of Physics: Conference Series **115** (2008) 012015 doi:10.1088/1742-6596/115/1/012015

vibrational excitation (VE) cross sections were recorded with a spectrometer using hemispherical electron-energy selectors [16]. The photoelectron(PE) spectra were recorded with a modified Perkin Elmer PS18 HeI photoelectron spectrometer.

Threshold energies for various fragmentations were calculated as the differences of the total energies of the products and the targets at 0 K, corrected for the zero point vibrational energy, using the density functional theory (DFT) B3LYP/6−311+G(2df,2p) model [17].

3. Results and Discussion

The spectra of the alcohols studied in this work – the two simplest alcohols (methanol and ethanol), one cyclic alcohol (cyclopentanol) and a pair of cyclic diols (*cis*-1,2-cyclopentadiol and *trans*-1,2-cyclopentanediol) – are shown in figures 1 and 2. The cyclic diols were chosen because of their relation to ribose, relevant for the study of DNA [18, 19].

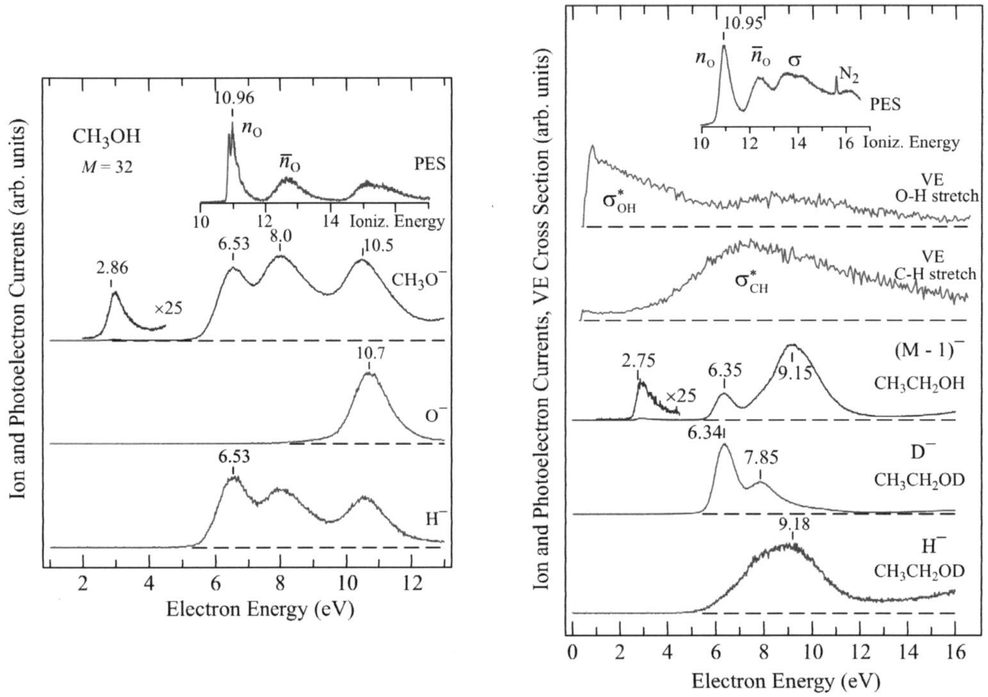

Figure 1. Yields of various fragments from methanol (left) and ethanol (right). PE spectra, shifted by −4.5 and −4.65 eV, are shown on the top. Vibrational excitation (VE) cross sections are shown for ethanol.

The spectra can be divided in two domains: low energy (0 - 5 eV) and high energy (5 - 10 eV). In the latter domain the chemistry, already well-known from smaller alcohols [20, 21], proceeds *via* Feshbach resonances, with a grandparent positive ion core and two electrons in diffuse Rydberg-like orbitals.

Additionally to the dissociations mediated by the Feshbach resonances, we report bands at lower energies, 2−3 eV. In all molecules the left sides of low energy bands are steep and consistent with vertical onsets at energetic thresholds (Table 1). Comparison of the threshold energies determined from experimental thermochemical data and calculated using quantum chemical methods for methanol and ethanol indicates that the latter are reliable within about ±0.2 eV, sufficient for the present purposes.

Table 1. Threshold energies for the reactions $e^- + \mathrm{ROH} \rightarrow \mathrm{RO}^- + \mathrm{H}$, in eV. The DFT values were calculated as described in section 2. The "therm." values were derived from experimental thermochemical data [22]. The onsets of the DEA bands are from the present spectra.

Target	E_{thr}		
	DFT	therm.	onset of DEA
methanol	2.82	2.94	2.86
ethanol	2.69	2.81	2.75
cyclopentanol	2.53		2.67
cis-cyclopentane-1,2-diol	1.89		1.94
trans-cyclopentane-1,2-diol	2.43		2.43

The low energy bands can not be assigned to core-excited resonances due to lack of suitable electronically excited parent state (either Rydberg or valence). At the same time the energy is too high for an assignment to a dipole-bound resonance or a virtual state. The dissociation must thus proceed *via* a $^2(\sigma^*)$ shape resonance.

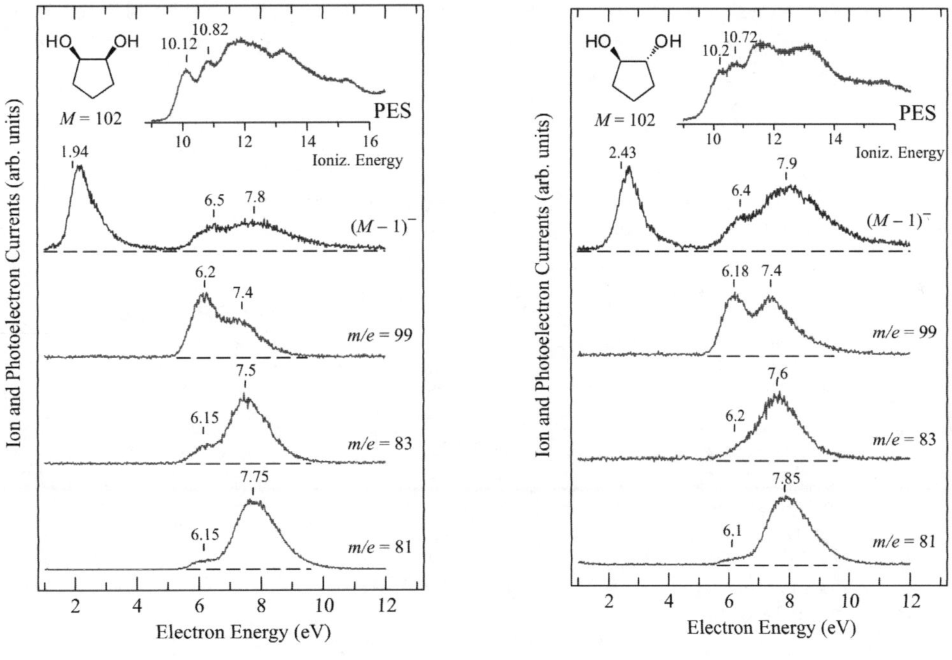

Figure 2. DEA spectra and PE spectra (shifted by −4.5 eV) of *cis*-cyclopentane-1,2-diol (left) and *trans*-cyclopentane-1,2-diol (right).

Additional evidence for the identity of resonance processes is provided, in the case of ethanol, by the cross sections for vibrational excitation (VE)(figure 1). The spectrum for the excitation of the C−H stretch vibration is similar to the C−H stretch excitation in hydrocarbons [23] – it consists of a very broad band peaking around 7.5 eV which could be assigned to a temporary occupation of σ^*_{CH} orbitals. The cross section for the excitation of the O−H stretch vibration is similar to the cross section observed in formic acid [24] – it peaks at threshold and then decreases slowly over an energy range of several eV. The electron cloud associated with this

Journal of Physics: Conference Series **115** (2008) 012015 doi:10.1088/1742-6596/115/1/012015

resonant process must be located on the O−H bond. The band in the O−H stretch vibrational excitation cross section is much broader than the band in the $(M-1)^-$ DEA band in Fig. 1, but we propose that both processes are due to the same σ^* shape resonance, possibly enhanced by dipole binding at low energies. The DEA bands at higher energies do not have any obvious analogs in VE, providing a further evidence that they are due to core excited Feshbach resonances and not to σ^* shape resonances.

The H^- yield is very similar in shape to the CH_3O^- yield above 6 eV, indicating that the same Feshbach resonances dissociate into both fragments – the charge distribution is not selective. The low energy band is missing in the H^- yield – probably because of the higher threshold for H^- formation.

Figure 2 shows the PES and DEA spectra of two representative diols. The oxygen lone pair orbitals give rise to two bands in the photoelectron spectrum. They are affected by a hydrogen bridge in the case of the *cis* compound while in the case of the *trans* compound they are subject to a destabilizing through-bond interaction, leading to the 'reverse order' of the cationic states [25].

The most striking feature of the DEA spectra of the diols is the decreasing energy and increasing relative intensity of the low energy band in comparison to the monoalcohols – they are more intense than the Feshbach resonance bands, in sharp contrast to the simple alcohols.

The decreasing thresholds are the results of the increased electron affinity of the fragments, particularly for the *cis* compound where the negative charge is stabilized by a hydrogen bridge. The increase in intensity could be a consequence of decreasing autodetachment width of the lowest σ^* resonance as a result of its decreased energy.

References

[1] Boudaiffa B, Cloutier P, Hunting D, Huels M A and Sanche L 2000 *Science* **287** 1658

[2] Martin F, Burrow P D, Cai Z, Cloutier P, Hunting D and Sanche L 2004 *Phys. Rev. Lett.* **93** 068101

[3] Schmelmer U, Jordan R, Geyer W, Eck W, Gölzhäuser A, Grunze M and Ulman A 2003 *Angew. Chem. Int. Ed.* **42** 559

[4] Abdoul-Carime H, Gohke S and Illenberger E 2004 *Phys. Rev. Lett.* **92** 168103

[5] Scheer A M, Możejko P, Gallup G A and Burrow P D 2007 *J. Chem. Phys.* **126** 174301

[6] König C, Kopyra J, Bald I and Illenberger E 2006 *Phys. Rev. Lett.* **97** 018105

[7] Flosadóttir H D, Denifl S, Zappa F, Wendt N, Mauracher A, Bacher A, Jónsson H, Märk T D, Scheier P and Ingólfsson O 2007 *Angew. Chem. Int. Ed.* **46** 8505

[8] Skalický T and Allan M 2004 *J. Phys. B* **37** 4849

[9] Ibănescu B C, May O, Monney A and Allan M 2007 *Phys. Chem. Chem. Phys.* **9** 3163

[10] Sanche L and Schulz G J 1971 *Phys. Rev. Lett.* **27** 1333

[11] Sanche L and Schulz G J 1973 *J. Chem. Phys.* **58** 479

[12] Gresteau F, Hall R I, Huetz A, Vichon D and Mazeau J 1979 *J. Phys. B* **12** 2925

[13] Spence D 1977 *J. Chem. Phys.* **66** 669

[14] Stepanović M, Pariat Y and Allan M 1999 *J. Chem. Phys.* **110** 11376

[15] Dressler R and Allan M 1985 *Chem. Phys.* **92** 449

[16] Allan M 1992 *J. Phys. B* **25** 1559

[17] Frisch *et al* M J Gaussian 03, Revision C.01 Gaussian, Inc., Wallingford, CT, 2004

[18] Aflatooni A, Scheer A M and Burrow P D 2006 *J. Chem. Phys.* **125** 054301

[19] Sulzer P, Ptasińska S, Zappa F, Mielewska B, Milosavljević A R, Scheier P, Märk T D, Bald I, Gohlke S, Huels M A and Illenberger E 2006 *J. Chem. Phys.* **125** 044304

[20] Kühn A, Fenzlaff H P and Illenberger E 1988 *J. Chem. Phys.* **88** 7453

[21] Curtis M G and Walker I C 1992 *J. Chem. Soc. Faraday Trans.* **88** 2805

[22] *NIST Chemistry webbook: http://webbook.nist.gov/chemistry*

[23] Walker I C, Stamatovic A and Wong S F 1978 *J. Chem. Phys.* **69** 5532

[24] Allan M 2006 *J. Phys. B* **39** 2939

[25] Hoffmann R, Heilbronner E and Gleiter R 1970 *J. Am. Chem. Soc.* **92** 706

15th International Symposium on Electron Molecule Collisions and Swarms IOP Publishing
Journal of Physics: Conference Series **115** (2008) 012016 doi:10.1088/1742-6596/115/1/012016

Rates for dissociative recombination of LiH^+ ions

R. Čurík[1] **and Chris H. Greene**[2]

[1] J. Heyrovský Institute of Physical Chemistry, Dolejškova 3, 18223 Prague 8, Czech Republic
[2] Department of Physics and JILA, University of Colorado, Boulder, CO 80309-0440, USA

E-mail: `roman.curik@jh-inst.cas.cz`

Abstract. We review recent progress in developing a theoretical treatment of the dissociative recombination (DR) process for the LiH^+ molecule, in which a low energy electron is captured and causes the molecule to dissociate into neutral fragments. This e+LiH^+ system is prototypical of the *indirect* class of DR processes, in which the incident electron destroys the molecule through Rydberg capture pathways. The conventional mechanism characteristic of most species with a high DR rate, which normally involves direct capture into a dissociative resonance potential curve, is entirely absent here. We have adopted *ab initio* multichannel quantum defect theory (MQDT) and a rovibrational frame transformation based on Siegert pseudostates to calculate indirect dissociative recombination for this simple diatomic ion. It is based on a set of *ab-initio* quantum defects as functions of the internuclear distance, which have been calculated using the R-matrix approach. The calculated DR rate coefficient is found to agree with recent experimental data [1].

1. Introduction

The diatomic molecule lithium hydride has served extensive duty as a test case for quantum chemistry methodologies. An online search for this tantalizingly simple molecule returns thousands of publications to date. One of the simplest chemical processes imaginable for this protoype species is the dissociative recombination (DR) reaction, which converts electronic energy into bond-breaking nuclear kinetic energy:

$$\mathrm{e} + \mathrm{LiH}^+ \rightarrow \mathrm{Li} + \mathrm{H} \quad (1)$$

The DR process for this simple species has seemed enigmatic because of two apparently conflicting facts: first, there is no curve crossing between the ground state of the ionic target LiH^+ and a resonance potential curve of the neutral LiH molecule that can provide a direct pathway from the incident electron scattering continuum into a dissociative Born-Oppenheimer state $\mathrm{Li}(nl\lambda)+\mathrm{H}(1s)$. Second, this molecule has a high measured DR rate coefficient [1], in the vicinity of $6(\pm 2) \times 10^{-7}$ cm^3/sec at $T = 139K$; when DR occurs this rapidly, it is normally believed to imply that a direct DR curve-crossing exists. On the other hand, this expectation has increasingly been recognized as an overgeneralization, because other ionic molecules exist, notably HeH^+ and H_3^+, where DR occurs rapidly via mediating Rydberg state captures.

Nevertheless, the first detailed theoretical study [2] of the dissociative recombination reaction for LiH^+ obtained a DR rate coefficient that was approximately 23 times lower than the lone storage ring experiment [1, 3, 4]. That discrepancy between a single theoretical result and a single experimental measurement stimulated our interest in studying this reaction using a different toolkit [5], which has found success in describing another system dominated by indirect Rydberg state captures, the H_3^+ molecule. The goal of this theoretical study [6, 7] has thus been two-fold: it would provide a test of the approximation scheme introduced in Ref.[5], and hopefully it would help to resolve the existing discrepancy and thereby shed light on the question of which (if either) of the theoretical and experimental results are quantitatively accurate. The components of that theoretical description relevant to a diatomic target ion include multichannel quantum defect theory (MQDT) [8], the rovibrational frame transformation [9, 10], and the use of Siegert pseudostates as the ionic vibrational channel functions in order to describe the actual dissociation of the complex into neutral Li+H fragments following the electron capture into a Rydberg state. The outcome of those calculations was very good agreement between our new theoretical results and the experiment; the largest remaining discrepancies occur only at sub-meV (parallel) collision energies, where our theoretical curve lies slightly below the experimental error bars.

We add that interest exists in the LiH^+ molecule because it participated in early universe chemistry and contributed to the cooling of primordial gas [11, 12, 13, 14]. As is the case with many molecular ions, the relative LiH^+, Li, H, and LiH abundances are controlled by photoionization, collisions, and by the dissociative recombination (DR) collisions with free electrons. This astrophysical interest makes it worth noting that even in the most recent detailed study of astrophysical processes that include LiH^+ DR, by Lepp *et al.* [15], a comparatively crude value for the DR rate coefficient was used in the modeling. Changing that rate coefficient from the value initially guessed by [11] to the experimentally measured one [3, 4] reduced the predicted abundance of LiH^+ for redshifts in the range $z < 25$ by approximately a factor of 3 in the latest model [15].

The purpose of the present article is to summarize our theoretical description, and to discuss the information that has been gleaned from this study and from the comparison between theory and experiment, as well as to present a few new results that shed light on other aspects of this system. For instance, we present the dependence of the DR rate on the initial rotational state of the ion, and new calculations of the dependence on the initial vibrational state.

2. Quantum Defects

In the present study we make use of the familiar MQDT theorem [8] that smoothly connects R-dependent quantum defects $\mu_{ll'}^{\Lambda}$ (multiplied by π) at energies just below the ionization threshold to R-dependent short-range scattering phase shifts just above the threshold:

$$\mu_{ll'}^{\Lambda}(R) = \frac{1}{\pi}\sum_{\gamma} X_{l\gamma}(R)\, \delta_{\gamma}^{\Lambda}(R)\, X_{l'\gamma}(R)\ , \tag{2}$$

where the $\delta_{\gamma}^{\Lambda}$ are the low-energy eigenphases for an e^- + LiH^+ collision, and the eigenvector matrix $\underline{X}(R)$ transforms the short-range K-matrix into diagonal form:

$$K_{ll'}^{\Lambda}(R) = \sum_{\gamma} X_{l\gamma}(R)\ \tan\delta_{\gamma}^{\Lambda}(R)\, X_{l'\gamma}(R)\ . \tag{3}$$

On the other hand Rydberg energy levels $U_{n\gamma}^{\Lambda}(R)$ of the neutral molecule LiH are described by eigenquantum defects $\mu_{\gamma}^{\Lambda}(R)$ that are eigenvalues of the $\mu_{ll'}^{\Lambda}(R)$ matrix as follows:

$$U^{+}(R) - U_{n\gamma}^{\Lambda}(R) = \frac{1}{2\left[n - \mu_{\gamma}^{\Lambda}(R)\right]^2}\ . \tag{4}$$

In this equation $U^+(R)$ is the potential curve of the LiH^+ ion and Λ denotes the projection of the Rydberg electron angular momentum l onto the axis of the diatomic molecule. We also

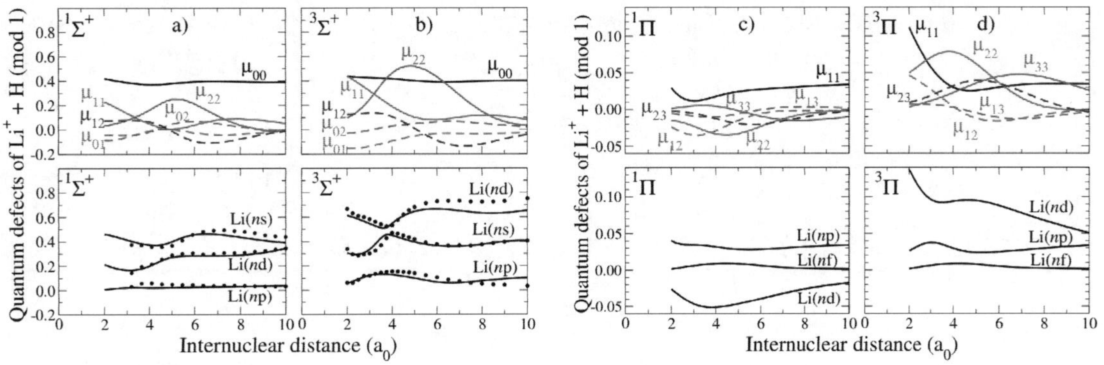

Figure 1. Dependence of calculated quantum-defect matrix elements on internuclear distance - Upper panels: a) $^1\Sigma^+$, b) $^3\Sigma^+$, c) $^1\Pi$, d) $^3\Pi$. Lower panels: The full curves are eigenvalues of the quantum defect matrices shown in the upper panels. The black dots are the n=4 eigenquantum defects extracted from the extensive CI calculations of Ref.[2].

introduced eigenindex γ distinguishing different Rydberg series of the LiH molecule as l is not a conserved quantum number.

In the previous theoretical study carried out by Florescu *et al* [2] the eigenquantum defects were obtained by use of Mulliken's formula (4) and the off-diagonal partial-wave couplings were neglected. However, in the present work we used the diatomic UK R-matrix package [16] to calculate the short-range K-matrix (3) by matching to coulomb functions at an R-matrix boundary of $r_0 = 25a_0$ and thus we were able to obtain all the partial-wave couplings as displayed in Fig. 1.

The lower panels of Fig. 1 also show the radial dependence of the calculated eigenquantum defects $\mu_\gamma^\Lambda(R)$, and demonstrate that they compare well with eigenquantum defects extracted from the $n = 4$ CI calculations carried out by Florescu *et al* [2]. This agreement is a numerical evidence for Seaton's theorem represented by eq.(2). Quantum defects calculated for $^1\Pi$ and $^3\Pi$ symmetries are displayed in Fig. 1 c) and d). Their values are an order of magnitude smaller than the Σ quantum defects, and for this reason they have negligible impact on the final DR results. Nevertheless, the small Π quantum defects can produce small shifts of individual resonances in the DR spectrum, and for this reason we include them in our calculations despite their minimal effect on the average DR rate.

3. Frame transformations

3.1. Vibrational frame transformation

The internuclear distance R is a good body frame "quantum number" when all electrons are confined within the box specified by r_0 and the Born-Oppenheimer approximation is accurate. The vibrational frame transformation connects R with the laboratory-frame quantization expressed by vibrational wave functions $\phi_{jv}(R)$. In the present study we have chosen a complete set of vibrational functions represented by Siegert pseudostates [17, 18]. They were obtained by solving the vibrational Schrödinger equation with the following boundary conditions:

$$\left[-\frac{d^2}{dR^2} + 2MU^+(R) + \frac{j(j+1)}{R^2} - k_{jv}^2\right]\phi_{jv}(R) = 0\,, \tag{5}$$

$$\phi_{jv}(0) = 0 \quad ; \quad \left(\frac{d}{dR} - ik_{jv}\right)\phi_{jv}(R)\Big|_{R_0} = 0\,. \tag{6}$$

In the above equations j is a rotational quantum number of the ion, M stands for its reduced mass, while R_0 denotes a nuclear radius beyond which we approximate the interaction potential in (5) to be constant. The standard vibrational frame transformation integral [20, 10] is modified by a surface term included because of the modified orthogonality relations of Siegert states [18, 23], yielding

$$S^{\Lambda}_{lv,l'v'}(j,j') = \int_0^{R_0} dR\, \phi_{jv}(R) \left(e^{2i\pi\underline{\mu}^{\Lambda}(R)}\right)_{l,l'} \phi_{j'v'}(R) + i\frac{\phi_{jv}(R_0)\left(e^{2i\pi\underline{\mu}^{\Lambda}(R_0)}\right)_{l,l'} \phi_{j'v'}(R_0)}{k_{jv}+k_{j'v'}} \,. \quad (7)$$

The underline in this equation denotes that $\underline{\mu}^{\Lambda}$ is a matrix with indices $\mu^{\Lambda}_{ll'}(R)$. Siegert pseudostate functions $\phi_{jv}(R)$ describe bound states and they also span over continuum space. We found all of the following results to be converged when the vibrational continuum is described by 20 outgoing-wave Siegert pseudostates for each j.

3.2. Rotational frame transformation

The rotational frame transformation transforms a set of body-frame angular functions (for further details see [21])

$$X^{l\Lambda\eta}_{JM}(\hat{r},\hat{R}) = \left(\frac{2J+1}{8\pi(1+\delta_{\Lambda 0})}\right)^{\frac{1}{2}} \left[Y_{l\Lambda}(\vartheta',\varphi')D^{J}_{\Lambda M}(0,\theta,\phi) + (-1)^{J}\eta Y_{l\Lambda}(\vartheta',\varphi')D^{J}_{\Lambda M}(0,\theta,\phi)\right] \quad (8)$$

to laboratory-frame angular functions

$$\Phi^{lj}_{JM}(\hat{r},\hat{R}) = \sum_m Y_{lm}(\vartheta,\varphi)Y_{jM-m}(\theta,\phi)(lm,jM-m|JM)\,. \quad (9)$$

In the above equations $\hat{R} \equiv (\theta,\phi)$ defines molecular axis orientation while $\hat{r} \equiv (\vartheta,\varphi) \equiv (\vartheta',\varphi')$ denotes a direction of scattered electron. Unprimed coordinates pertain to the laboratory frame and the primed ones to the body frame. The total angular momentum J is defined as a sum of molecular rotational quantum number j and electron angular momentum l : $\boldsymbol{J} = \boldsymbol{l} + \boldsymbol{j}$. It is important to mention that in eqs. (8),(9) it has already been assumed that the target molecule is in a Σ^+ state. Both wave functions have definite parity η under inversion I of all coordinates:

$$IX^{l\Lambda\eta}_{JM} = \eta X^{l\Lambda\eta}_{JM}, \quad (10)$$

$$I\Phi^{lj}_{JM} = (-1)^{l+j}\Phi^{lj}_{JM}. \quad (11)$$

The LiH^+ ion is treated in Hund's case (b), with spin-orbit coupling neglected. The eigenstates possess a definite inversion parity, namely $\eta = (-1)^{l+j}$. This procedure transforms the body frame scattering matrix (7) into the laboratory frame, as follows:

$$S^{J\eta}_{ljv,l'j'v'} = \sum_{\Lambda} U^{\Lambda\eta}_{Jjl}\, S^{\Lambda}_{vl,v'l'}(j,j')\, U^{\Lambda\eta}_{Jj'l'}\,. \quad (12)$$

The real, orthogonal rotational transformation matrix $U^{\Lambda\eta}_{Jjl}$ is taken from [21].

4. Dissociative recombination rates

The scattering matrix $\underline{S}$ on the left-hand side of Eq.(12) represents an amplitude for electron-ion scattering from an initial channel defined by $(l'j'v')$ into a final channel defined by (ljv). However at a particular collision energy not all of the final channels are typically open. Complicated Rydberg structure associated with closed-channels resonances is generated by the "elimination of closed channels" formula [22] that reduces the unphysical S matrix (12) down to the physically-observable open-channel subspace:

$$\underline{S}^{\rm phys} = \underline{S}^{\rm oo} - \underline{S}^{\rm oc}\left[\underline{S}^{\rm cc} - e^{-2i\underline{\beta}(E)}\right]^{-1}\underline{S}^{\rm co}\,. \tag{13}$$

The superscripts o and c respectively denote open- and closed-channel sub-blocks of the unphysical S-matrix (12), while $\underline{\beta}(E)$ is a diagonal matrix of effective Rydberg quantum numbers with respect to the closed-channel thresholds. Because the high ionization thresholds are described by a Siegert pseudo-continuum state with finite width, $\underline{\beta}(E)$ becomes complex. This fact results in the $\underline{S}^{\rm phys}$ matrix becoming sub-unitary. Physically, the lost flux is associated with an electronically closed channel attached to a vibrational wave function described by a dissociative outgoing-wave Siegert state. The departure from unitarity was identified by [23, 5, 6] as the dissociation probability following electron impact in the specified incident channel $i' \equiv (v'j'l')$:

$$\sigma_{i'}^{J\eta}(\varepsilon_{i'}) = \frac{\pi}{2\varepsilon_{i'}}\left[1 - \sum_i S_{ii'}^{\rm phys}(E)\, S_{i'i}^{\dagger\,\rm phys}(E)\right]\,, \tag{14}$$

with the incident electron collision energy $\varepsilon_{i'} = E - E_{j'v'}$. The experimentally observable cross-section for dissociation following electron impact is then

$$\sigma_{j'v'}(\varepsilon_{j'v'}) = \frac{1}{2j'+1}\sum_{\eta J l'}(2J+1)\,\sigma_{l'j'v'}^{J\eta}(\varepsilon_{j'v'})\,. \tag{15}$$

The corresponding DR rate is then obtained after multiplying the cross-section by the incident electron velocity.

This calculated DR rate exhibits an infinite number of resonances accumulating near each closed-channel ionization threshold. They are also called vibrational (or rotational) Feshbach resonances, and they represent autoionizing and predissociating states of LiH. In all practical applications, the raw DR rate obtained in this manner is averaged over the relevant energy and eigenstate distributions of the ions and of the colliding electrons. In the following we discuss two important cases:

i) In the storage-ring experiments as [1], the electrons and ions are not in thermal equilibrium. The temperature of the electron beam is usually lower than the temperature of the ions. Furthermore, the momentum spread of the electrons is anisotropic. The parallel energy spread $\Delta E_\parallel$ is very well controlled, while the distribution is much broader in the perpendicular components of the relative velocity vector. ($\Delta E_\parallel = 0.1$ meV and $\Delta E_\perp = 12$ meV in [1]). The convolution over parallel and perpendicular energy distributions has been performed as was outlined in [5] and elaborated in detail in [24]. Figure 2 summarizes our results along with previous experimental and theoretical results. This figure also demonstrates the results of a numerical test conducted to interpret the discrepancy between our theoretical results and those of [2]. Specifically, we have performed one set of calculations that neglect the off-diagonal l-mixing to mimic the calculations performed by [2], i.e. using only the diagonal eigenvalue form $\mu_\gamma^\Lambda(R)$ of quantum-defect matrix (shown in fig. 1 a,b). Introduction of this approximation lowers the DR rate by an order of magnitude, and this artificially restricted calculation agrees with the results of [2].

ii) In the interstellar medium we assume that the electrons are distributed isotropically in momentum space. Thus the thermal rate coefficient is obtained by averaging the DR rate over a Maxwellian distribution of electrons at a temperature T:

$$\alpha_{th}(j',v',T) = \frac{8\pi}{(2\pi kT)^{3/2}}\int_0^\infty dE\,\exp\left(-\frac{E}{kT}\right)\sqrt{\frac{E}{2}}\,\alpha_{j',v'}(E)\,. \tag{16}$$

In Fig. 3 we show the calculated thermal rate coefficient α_{th} as a function of the temperature T of electrons for a number of ionic initial states (j', v'). Our calculations are compared with the estimates used in models of lithium chemistry of the early universe [11, 15] (thick full and dashed curves in the right panel). Drops of the thermal DR rates together with some small oscillations under 10K visible in the Fig. 3 are caused by the fact that the energy distributions of such cold electrons are narrow enough to distinguish vibrational Feshbach resonances occuring close to zero collision energy. In other words, with different initial rotational level the zero collision energy threshold moves through the dense structure of the closed-channel resonances resulting in an oscillatory behavior of the DR rate at very low temperatures.

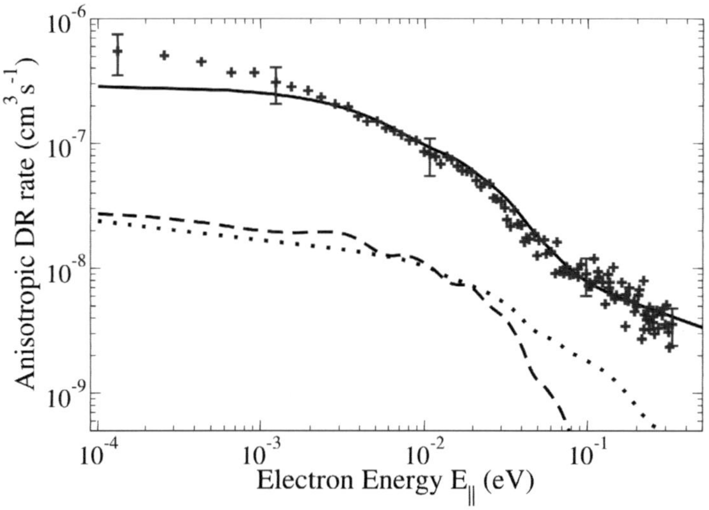

Figure 2. DR rate: The solid curve is our calculated, anisotropically-averaged rate for $\Delta E_{\parallel} =$ 0.1 meV and $\Delta E_{\perp} = 12$ meV. The broken curve shows the calculation of [2]. The dotted curve is our truncated result obtained by neglecting the off-diagonal couplings in the quantum-defect matrix (see the fig. 1). The crosses denote the experimental data [3], with a few characteristic error bars shown.

5. Summary and Conclusions

We applied ro-vibrational frame-transformation techniques to calculate indirect dissociative recombination of LiH^+ ions. The short-range scattering matrices and quantum defects necessary for the technique were calculated by the fully *ab-initio* R-matrix method. The diagonal form of the quantum defect matrix agrees very well with the eigenquantum defects extracted from standard quantum chemistry calculations of Florescu *at al* [2].

Encouraged by the very good agreement between our calculated anisotropically-averaged DR rate and the storage rings experiments [1, 3] we also calculated isotropic thermal rate coefficient used in models of lithium chemistry in the early universe [11, 15]. We explored the dependence of the thermal rate coefficients on the initial ro-vibraional state of the LiH^+ ion. As the higher vibrational states are more strongly coupled to the continuum it is not surprising that the dissociation is faster for the vibrationally hotter cations.(fig. 3, left panel) However, we have found very weak dependence of the DR rate on the initial rotational state of the molecule with the temeperature of the electrons larger than 20 K. For lower temperatures the energy distribution

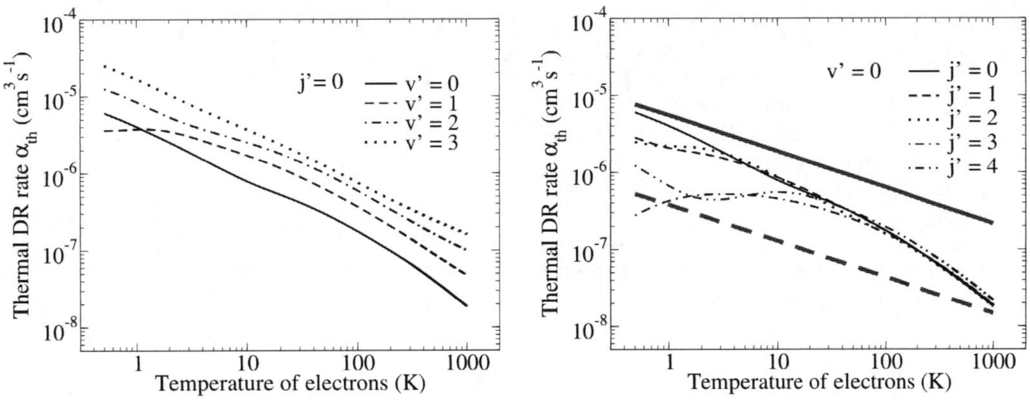

Figure 3. Dependence of the thermal rate α_{th} on the electron temperature and on the initial state of the ion: Left panel: Dependence on the initial vibrational state, with the initial rotational state fixed as $j' = 0$. Right panel: Dependence on the initial rotational state, with the initial vibrational state fixed as $v' = 0$. The thick dashed line is the theoretical estimate utilized by [11] while the thick full line displays the later correction that was published in [15].

window is narrow enough to distinguish different close-channel resonances occuring near zero-energy threshold and oscillatory structures appear (fig. 3, right panel). Our results suggest that rates used in some models of lithium chemistry in the early universe [11] are underestimated by an order of magnitude. The more recent corrected values used in astrophysical modeling [15], on the other hand, appear to be overestimated at temperatures higher than a few Kelvin, by as much as a factor of 4 at 1000K.

Acknowledgments

This work was supported in part by NSF grants CHE-0446688, OISE-0532040, ITR-0427376, the project (EIPAM) sponsored by the European Science Foundation (grant No. PESC7-20), the Academy of Sciences of the Czech Republic (grants No. A100400501, KJB400400803) and the Grant Agency of the Czech Republic (grant No. 202/08/0631). We thank S. Tonzani for assistance and discussions in the early stages of the project. The work of C.H.G. was also supported in part by the Alexander von Humboldt Foundation, and by the Max-Planck-Institute for Nuclear Physics in Heidelberg, Germany.

References

[1] S. Krohn,*et al.*, *Phys. Rev. Lett.* **86**, 4005 (2001)
[2] A. I. Florescu, A. Suzor-Weiner, T. Leininger, F. X. Gadéa, *Phys. Scrip.* **T110**, 172 (2004)
[3] S. Krohn, *PhD thesis at the Combined Faculties for the Natural Sciences and Mathematics of the Ruperto-Carola University of Heidelberg, Germany*, 102 (2001)
[4] D. Zajfman, D. Schwalm and. A Wolf, *Hyperfine Interactions* **146**, 265 (2003)
[5] V. Kokoouline, C. H. Greene, *Phys. Rev. A* **68**, 012703 (2003); see also S. Fonseca dos Santos, V. Koko ouline, and C. H. Greene, *J. Chem. Phys.* **127**, 124309 (2007)
[6] R. Čurík and C. H. Greene, *Phys. Rev. Lett.* **98**, 173201 (2007)
[7] R. Čurík and C. H. Greene, *Mol. Phys.* **105**, 1565 (2007)
[8] M. J. Seaton, *Rep. Prog. Phys.* **46**, 167 (1983)
[9] U. Fano, *Phys. Rev. A* **2**, 353 (1970)
[10] C. H. Greene, Ch. Jungen, *Phys. Rev. Lett.* **55**, 1066 (1985)
[11] P. C. Stancil, S. Lepp, A. Dalgarno, *ApJ* **458**, 401 (1996)
[12] A. Dalgarno, K. Kirby, P. C. Stancil, *ApJ* **458**, 397 (1996)
[13] F. A. Gianturco, P. G. Giorgi, *ApJ* **479**, 560 (1997)

[14] E. Bodo, F. A. Gianturco, R. Martinazzo, *Phys. Rep.* **384** 85 (2003)
[15] S. Lepp, P. C. Stancil, A. Dalgarno, *J. Phys. B: At. Mol. Opt. Phys.* **35**, R57 (2002)
[16] L. A. Morgan, C. J. Gillan, J. Tennyson, X. Chen, *J. Phys. B.* **30**, 4087 (1997)
[17] A. J. F. Siegert, *Phys. Rev.* **56**, 750 (1939)
[18] O. I. Tolstikhin, V. N. Ostrovsky, H. Nakamura, *Phys. Rev. A* **58**, 2077 (1998)
[19] H. Berriche, F. X. Gadé a, *Chem. Phys.* **191**, 119 (1995)
[20] Ch. Jungen, D. Dill, *J. Chem. Phys.* **73**, 3338 (1980)
[21] E. S. Chang, U. Fano, *Phys. Rev. A* **6**, 173 (1972)
[22] M. Aymar, C. H. Greene, E. Luc-Koenig, *Rev. Mod. Phys.* **68**, 1015 (1996)
[23] E. L. Hamilton, C. H. Greene, *Phys. Rev. Lett.* **89**, 263003 (2002)
[24] V. Kokoouline, C. H. Greene, *J. Phys.: Conf. Ser. 4*, 74 (2005)

Journal of Physics: Conference Series **115** (2008) 012017 doi:10.1088/1742-6596/115/1/012017

Relaxation processes of electrons and positrons in gases in electric and magnetic fields crossed at arbitrary angles

S Dujko[1,2], R D White[1], K F Ness[1], R E Robson[3] and Z Lj Petrović[2]

[1]ARC Centre for Antimatter-Matter Studies, School of Mathematics, Physics and IT, James Cook University, Townsville 4811, Australia

[2]Institute of Physics Belgrade, Pregrevica 118,11080 Belgrade, Serbia

[3]ARC Centre for Antimatter-Matter Studies, Research School of Physics and Engineering, Australian National University, Canberra 2600, Australia

E-mail: sasa.dujko@jcu.edu.au

Abstract. A multi-term solution of the Boltzmann equation has been developed and used to investigate the temporal relaxation of charged-particle swarms under the influence of electric and magnetic fields crossed at arbitrary angles. A Monte Carlo simulation technique has been used to verify the Boltzmann equation analysis under hydrodynamic conditions and for an independent study of spatial relaxation of the electrons under non-hydrodynamic conditions. We present results for model and real gases highlighting the explicit influence of the magnetic field and non-conservative collisions on temporal and spatial relaxation characteristics, including the existence of transiently negative electron diffusivity. We present results for thermalization of the positrons in nitrogen highlighting the applicability of our theory for contemporary kinetic studies in the domain of positron physics.

1. Introduction

The theoretical investigation of charged-particle swarms moving in an unbounded gas in electric and magnetic fields is a topic of great interest both as a problem in basic physics and for its potential for application to modern technology. One of the major challenges in these investigations is an accurate representation of temporal and spatial relaxation of charged-particle swarms. Relaxation processes of a swarm of charged particles are related to various problems of gaseous electronics such as modeling of non-equilibrium plasma discharges, high-speed switching technique, swarm physics and physics of gas lasers [1]. The knowledge of temporal relaxation is essential for a better understanding of electron-molecule interaction as well as for a better understanding of transient transport phenomena in gases such as transient negative electron mobility [2] or transient negative electron diffusivity [3]. In the context of plasma modeling, the relaxation times of the electron transport properties are necessary input data for fluid plasma models such as the relaxation continuum model [4]. In addition, to fully appreciate the complex structure of the transport properties in radio-frequency (rf) electric and magnetic fields [5,6], a systematic investigation of the temporal relaxation of a swarm of charged particles in dc electric and magnetic fields is required.

While the studies of temporal relaxation of the electrons are tremendously important from the point of view of fundamental physics, modern non-hydrodynamic studies of spatial relaxation of the electrons are motivated by the need to understand the electron kinetics in low-temperature plasmas, especially in the neighborhood of sources and boundaries. A number of theoretical methods to calculate the electron transport properties under these "non-hydrodynamic" conditions, especially those attempting to model or simulate the cathode fall of a dc [7] or sheath of rf discharges [8], revealed the non-local nature of electron transport. One of the principal aims of this paper is to study the non-local nature of electron transport when a magnetic field is present and non-conservative collisions are operative.

We begin this paper with a brief review of multi-term theory for solving the Boltzmann equation valid for both electrons and ions in time-dependent electric and magnetic fields crossed at arbitrary angles, incorporating the effects of non-conservative collisional processes. We focus on two situations: (i) temporal relaxation of the electrons and (ii) thermalization of the positrons. In addition to the Boltzmann equation analysis, we discuss our hydrodynamic and non-hydrodynamic Monte Carlo simulation codes. The former is used to test the Boltzmann equation analysis under the hydrodynamic conditions. The latter is used for studies of the effects of magnetic fields and non-conservative collisions on spatial relaxation of the electrons. After giving brief reviews of theoretical methods, we then give numerical examples for various cases of special interest, highlighting recent new results.

2. Theory

2.1. Multi-term theory for solving the Boltzmann equation

The behaviour of charged particles in gases under the influence of electric and magnetic fields is described by the phase-space distribution function $f(\mathbf{r},\mathbf{c},t)$ representing the solution of the Boltzmann equation

$$\frac{\partial f}{\partial t}+\mathbf{c}\cdot\frac{\partial f}{\partial \mathbf{r}}+\frac{e}{m}[\mathbf{E}+\mathbf{c}\times\mathbf{B}]\cdot\frac{\partial f}{\partial \mathbf{c}}=-J(f,f_0), \tag{1}$$

where $\mathbf{r}$ and $\mathbf{c}$ denote the position and velocity co-ordinates, e and m are the charge and mass of the swarm particle and t is time. The electric and magnetic fields are assumed to be spatially homogeneous with magnitudes $\mathbf{E}$ and $\mathbf{B}$ respectively. In what follows, we employ a co-ordinate system in which $\mathbf{E}$ defines the z-axis while $\mathbf{B}$ lies in the y-z plane, making an angle ψ with respect to the $\mathbf{E}$. Swarm conditions are assumed to apply where the charged particle number density is much less than number density of neutral species and mutual interactions between swarm particles are negligible compared with swarm particle-neutral particle interactions. The right-hand side of Eq. (1) denotes the linear charged particle-neutral molecule collision operator, accounting for elastic, inelastic and non-conservative (e.g. ionizing and attaching) collisions. For elastic collisions we use the original Boltzmann collision operator [9], while for inelastic collisions we prefer the semiclassical generalization of Wang-Chang et al. [10]. The attachment and ionization collision operators employed are detailed in Ref. [11].

(a) Spherical-harmonic expansion. The directional dependence of the phase-space distribution function in velocity space is represented in terms of a spherical harmonic expansion

$$f(\mathbf{r},\mathbf{c},t)=\sum_{l=0}^{l_{\max}}\sum_{m=-l}^{l} f_m^{(l)}(\mathbf{r},c,t)Y_m^{[l]}(\hat{\mathbf{c}}), \tag{2}$$

where $\hat{\mathbf{c}}$ represents the angles of $\mathbf{c}$. The value of l_{max} is incremented until some predefined accuracy criterion is satisfied. This value indicates the deviation of the velocity distribution function from isotropy.

(b) Density gradient expansion. Assuming the hydrodynamic regime, the spatial dependence is represented by

$$f_m^{(l)}(\mathbf{r},c,t)=\sum_{s=0}^{2}\sum_{\lambda=0}^{s}\sum_{\mu=-\lambda}^{\lambda} f(lm\mid s\lambda\mu;c,t)G_\mu^{(s\lambda)}n(\mathbf{r},t)\,, \tag{3}$$

where $G_\mu^{(s\lambda)}$ is the irreducible gradient operator [11]. Truncation at s = 2 is necessary to determine transport coefficients up to and including diffusion when non-conservative collisions are operative.

(c) Sonine polynomial expansion. The speed dependence of the coefficients in Eq. (3) is represented by an expansion about a Maxwellian at an arbitrary temperature T_b, in terms of Sonine polynomials

$$f(lm\mid s\lambda\mu;c,t)=w(\alpha,t)\sum_{\nu=0}^{\infty}F(\nu lm\mid s\lambda\mu;\alpha,t)R_{\nu l}(\alpha c)\,, \tag{4}$$

where

$$R_{\nu l}(\alpha c)=N_{\nu l}\left[\frac{\alpha c}{\sqrt{2}}\right]^l S_{l+1/2}^{(\nu)}\left(\frac{\alpha^2c^2}{2}\right), \tag{5}$$

$$w(\alpha,c)=\left[\frac{\alpha^2}{2\pi}\right]^{3/2}\exp\left\{-\frac{\alpha^2c^2}{2}\right\}, \tag{6}$$

$$\alpha^2=\frac{m}{kT_b}, \tag{7}$$

$$N_{\nu l}^2=\frac{2\pi^{3/2}\nu!}{\Gamma(\nu+l+3/2)}, \tag{8}$$

and $S_{l+1/2}^{(\nu)}\left(\frac{\alpha^2c^2}{2}\right)$ are Sonine polynomials. Using the appropriate orthogonality relations of the spherical harmonics and modified Sonine polynomials the following system of coupled differential equations for the moments $F(\nu lm\mid s\lambda\mu;\alpha,t)$ is generated:

$$\sum_{\nu'=0}^{\infty}\sum_{l'=0}^{\infty}\sum_{m'=-l'}^{l'}\Big[\partial_t\delta_{\nu\nu'}\delta_{ll'}\delta_{mm'}+n_0J_{\nu\nu'}^{l}\delta_{ll'}\delta_{mm'}+\omega(000)\delta_{\nu\nu'}\delta_{ll'}\delta_{mm'}$$
$$+i\frac{q}{m}E(lm\mid l'm10)\alpha_t\left\langle \nu l\middle\|K^{[1]}\middle\|\nu'l'\right\rangle\delta_{mm'}$$
$$+\frac{qB}{m}\left\{\frac{\sin\psi}{\sqrt{2}}\Big[\sqrt{(l-m)(l+m+1)}\delta_{m'm+1}-\sqrt{(l+m)(l-m+1)}\delta_{m'm-1}\Big]-im\cos\psi\delta_{m'm}\right\}\delta_{\nu\nu'}\delta_{ll'}$$
$$-n_0J_{0\nu'}^{0}F(\nu lm\mid 000)(1-\delta_{s0}\delta_{\lambda0}\delta_{\mu0})\delta_{l'0}\delta_{m'0}\Big]F(\nu'l'm'\mid s\lambda\mu)=X(\nu lm\mid s\lambda\mu), \tag{9}$$

where $J_{\nu\nu'}^{l}$ and $\left\langle \nu l\middle\|K^{[1]}\middle\|\nu'l'\right\rangle$ are reduced matrix elements, $\omega(000)$ represents the net creation rate and $(lm\mid l'm10)$ is a Clebsch-Gordan coefficient. The explicit expressions for the RHS are given in Ref. [12] for the case of crossed dc electric and magnetic fields. These expressions are modified in this paper in order to consider the time-dependent fields crossed at arbitrary angles in the presence of non-conservative collisions. Discretising in time using an implicit finite difference scheme converts the system of coupled differential equations into a hierarchy of coupled matrix equations. To establish the transport coefficients of interest we are required to solve the following members of the hierarchy (s,λ,μ) = (0,0,0), (1,1,0), (1,1,1), (2,0,0), (2,2,0), (2,2,2). The explicit expressions for both bulk and flux transport coefficients including the mean energy and gradient energy vector components are given in Ref. [6]. Transport coefficients of interest in this work include the mean energy (ε), drift velocity components (W_x, W_y, W_z) and diagonal diffusion tensor elements (D_{xx}, D_{yy}, D_{zz}).

2.2. Monte Carlo simulation technique

The other method for investigating the transient effects of charged-particle transport in dc electric and magnetic fields is a Monte Carlo simulation technique. In the context of hydrodynamic studies, it is assumed that the charged-particle swarm develops in an infinite space. At time $t = 0$, electrons are initially released from the origin according to Maxwellian velocity distribution with the mean starting energy of 1 eV under the influence of electric field only. After relaxation to the steady state in which both the swarm transport parameters and distribution function do not change in time, the magnetic field is applied and the relaxation process is followed accurately in time. An extremely large number of electrons (typically 10^7) have been followed in a neutral gas in order to obtain good statistics of the output data, particularly the diffusion coefficients. It is assumed that the electron density is sufficiently small so that Coulomb interaction between the electrons as well as shielding of the filed is negligible. All calculations are performed at zero gas temperature.

Spatial relaxation of the electrons has been studied by our steady-state Townsend Monte Carlo simulation code. The primary electrons are isotropically released one by one from the cathode surface into the half space with an initial energy ε_0. When an ionization collision occurs, a set of all dynamic properties (the moment of an ionization collision, the position of new electron, the starting energy and velocity) of secondary electron are established and placed at the stack. When primary electron reaches the anode surface or disappears in an attachment collision event, the first available electron from the stack is followed. These secondary electrons from the stack are released isotropically. In an attachment collision the electron is consumed and hence not followed further. If the stack is empty, the next primary electron is released and the whole procedure repeats. Thermal motion of the background neutral particles and electron-electron interactions are neglected. The electrodes are considered to be perfectly absorbing.

In both the hydrodynamic and non-hydrodynamic versions of Monte Carlo codes, we follow the spatiotemporal evolution of each electron through time steps governed by the minimum of two relevant time constants: mean collision time and cyclotron period for **E**×**B** field. These finite time steps are used to solve the integral equation for the collision probability in order to determine the time of the next collision. Once the moment of the next collision is established, the nature of the collision is determined by using the relative probabilities of the various collision types. All electron scattering is assumed to be isotropic regardless of the collision nature.

The way of sampling of transport properties depends upon the manner of swarm observation [13]. The definitions and corresponding formulae for the electron transport coefficients under the hydrodynamic conditions were given in our previous publications [14]. The spatially resolved electron transport properties have been obtained using the following formula [13]:

$$\langle \xi \rangle_j = \frac{\frac{1}{\Delta z}\int_{z_j-\Delta z/2}^{z_j+\Delta z/2} \xi f_{SST}(z,\mathbf{v})\,dz\,dv}{\frac{1}{\Delta z}\int_{z_j-\Delta z/2}^{z_j+\Delta z/2} f_{SST}(z,\mathbf{v})\,dz\,dv} \approx \frac{\sum_{k=1}^{N} \xi_k^j \Delta t_k^j}{\sum_{k=1}^{N} \Delta t_k^j}, \tag{10}$$

where $f_{SST}(z,\mathbf{v})$ is the steady state distribution function, ξ_k^j is the value of the quantity to be sampled when kth electron is contained in jth box, Δt_k^j is the residence time of the electron in that box and N is the total number of electrons which appear there. Electrons moving towards both the cathode and anode must be considered and sampled.

3. Results and discussion

3.1. Temporal relaxation of the electrons

In this section we consider the response of the electron transport coefficients to the application of a magnetic field under hydrodynamic conditions. Similar studies have been published previously for neon using the two-term approximation [15]. We extend these studies by (i) overcoming the inherent inaccuracies of the two-term approximation and (ii) addressing the temporal relaxation of spatial inhomogenities through a study of the diffusion tensor. In addition, this work is an extension of the work in [16] to consider the explicit influence of non-conservative collisions and the angle between the fields on the temporal relaxation profiles. Two specific situations have been considered: (i) the conservative Reid ramp model gas system [17]; and (ii) the ionization model gas of Lucas and Saelee [18]. The former has become the standard test for Monte Carlo simulations and Boltzmann equation solutions, particularly in light of known failure of the two-term approximation for this model. The latter is significant for demonstrating just how important it is to consider the ionization as a true non-particle conserving collision, and not simply just another inelastic process, as is sometimes done [21]. The initial conditions represent the steady state magnetic field free case where the electron swarm is acted on solely by a dc electric field (E/n_0 = 12 Td; B/n_0 = 0 Hx; 1 Td = 10^{-21} Vm^2; 1Hx = 10^{-27} Tm^3). At time t = 0, a magnetic field is switched on and the relaxation properties are monitored as a function of time (normalized time n_0t).

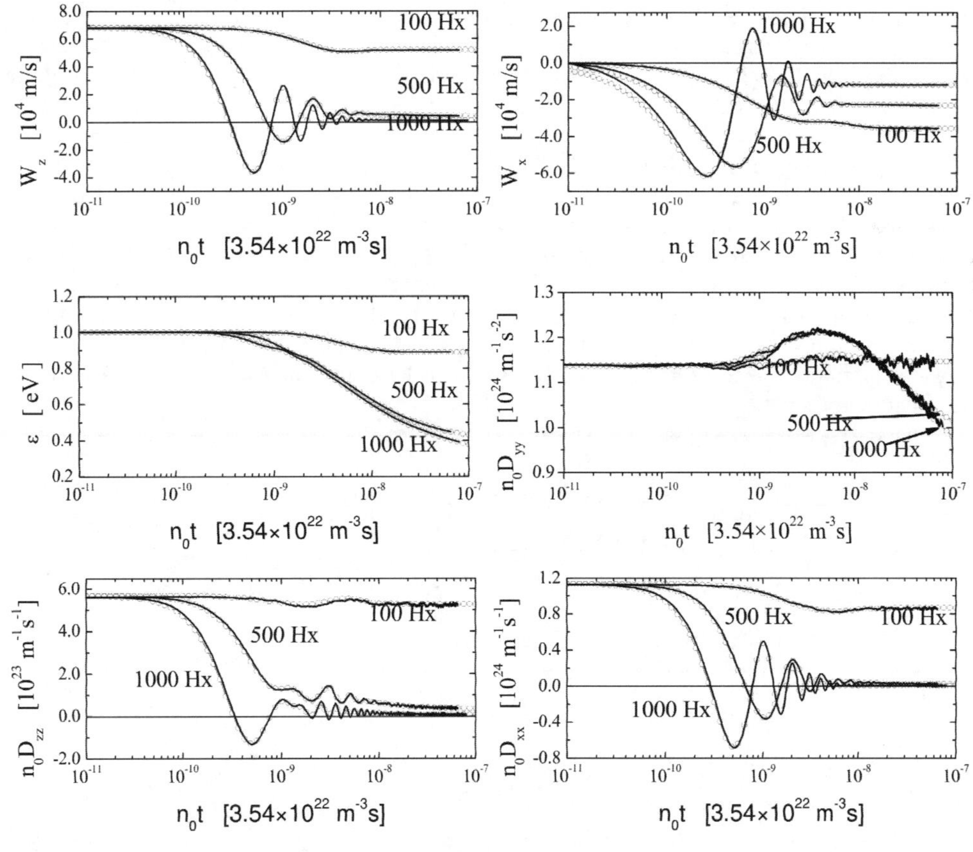

Figure 1. Temporal relaxation of the mean energy, drift velocity components and diagonal elements of the diffusion tensor for various applied magnetic fields in a crossed field configuration for electrons for Reid ramp model (solid lines-Monte Carlo; open symbols-Multi-term Boltzmann equation).

In Fig.1 we show the temporal relaxation profiles of the electron transport coefficients for the Reid ramp model for various B/n_0 in a crossed field configuration. The results obtained by multi-term theory for solving the Bolztmann equation are compared with those obtained by Monte Carlo simulation technique. The excellent agreement between these methods validates the basis of transport theory as well as numerical integrity of both numerical codes. While the mean energy and B-component of the diffusion tensor show monotonic relaxation profile, the relaxation profiles of the drift and diffusion in the **E** and **E**×**B** direction exhibit a transition from monotonic decay to damped periodic decay as magnetic field increases. For these damped periodic profiles, the oscillations are on the time-scale of the gyro-orbits τ and the envelope decays on a time-scale for momentum relaxation τ_m together with a further relaxation on the time-scale of the energy relaxation time τ_e. Perhaps the most striking phenomena is the existence of transiently negative excursions of the diffusion tensor elements in both the **E** and **E**×**B** directions. The existence of transiently negative diagonal diffusion elements in swarms was observed in CO_2 [16] as well as in rf fields [19] and is explained in greater detail in [3].

In Fig. 2 we show the temporal relaxation profiles of D_{yy} as a function of the magnetic field strengths and angle between the fields. The results obtained by the multi-term theory (MT) are compared with those obtained by the two-term approximation (TTA) for solving the Boltzmann equation. In the early and intermediate stages of the relaxation process, the significant deviations between the temporal profiles can be observed. This is a clear indication that the initial distribution function and its initial evolution deviates substantially from isotropy in velocity space. However, as magnetic field and the angle between the fields increase, the deviations between TTA and MT results are significantly diminished. This suggests that the magnetic field acts to destroy the anisotropy of the velocity distribution function, consequently inducing enhanced convergence in the l-index. A similar effect was observed in methane [20].

Another striking property is the high sensitivity of the relaxation profiles to the angle between the fields. In contrast to other diagonal elements of the diffusion tensor, the oscillatory nature of D_{yy} is reduced as the angle between the fields is increased. Finally, in the limit of an orthogonal field configuration these profiles are monotonic (see Fig. 1). For an orthogonal field configuration the Lorentz force does not act in this direction and hence there are no imprinted oscillations on the diffusion coefficient in this direction. On the other hand, for small angles between the fields, the electrons are under the action of Lorentz force producing the oscillatory relaxation profiles. For a magnetic field of 1000 Hx and angle of 30° between the fields, the Lorentz force is shown to induce the negative transient diffusivity.

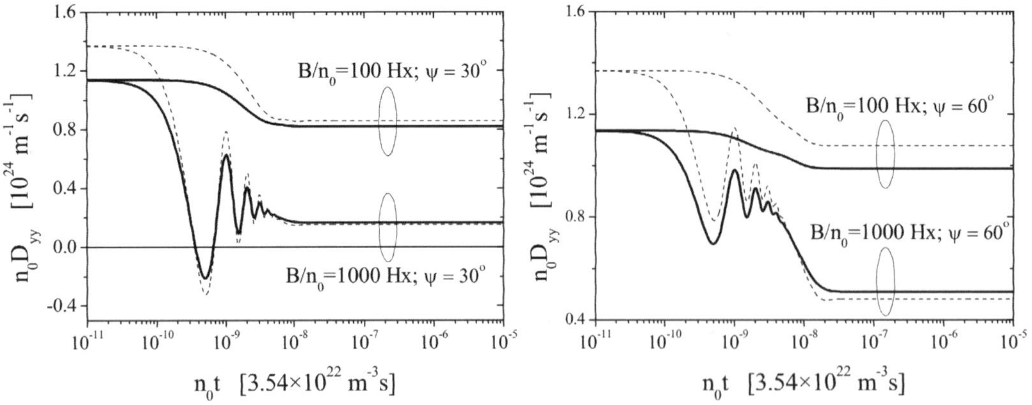

Figure 2. Temporal relaxation of D_{yy} for magnetic fields strengths of 100 and 1000 Hx and angles of 30 and 60 degrees between the fields for Reid ramp model (solid lines-MT; dashed lines-TTA)

In Fig. 3 we demonstrate the influence of non-conservative collisions on temporal relaxation of W_y and W_z for Lucas Saelee ionization model. We show the bulk and flux values of the drift velocity components as a function of various magnetic fields and angles between the fields. The parameter F controls the ionization degree. In the early stage of the relaxation process there is no distinction between the bulk and flux values of W_y indicating there is no spatial variation of the average electron energy in this direction. On the other hand, there is a clear distinction between the bulk and flux values for W_z. This suggests a strong variation of the average energy along the swarm in this direction. However, as the relaxation process proceeds in time, the synergism of magnetic field and ionization collisions on the temporal relaxation profiles becomes clearly evident. While the distinction between the bulk and flux values for W_y is more pronounced, the distinction for W_z is firstly diminished at the intermediate stage of a relaxation process and then upon reaching the steady state there are no differences between the bulk and flux values. These results suggest that due to complex interplay of the action of the magnetic field and the energy and momentum dissipation of the electrons in collisions, a complicated redistribution of high energetic electrons occurs.

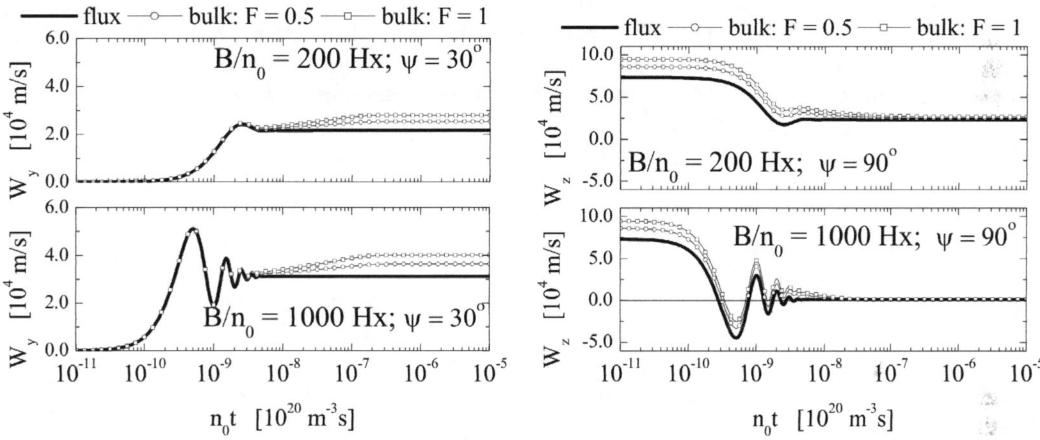

Figure 3. Temporal relaxation of the bulk and flux drift velocity components W_y and W_z for magnetic fields strengths of 200 and 1000 Hx and angles of 30 and 60 degrees between the fields for the ionization model of Lucas and Saelee.

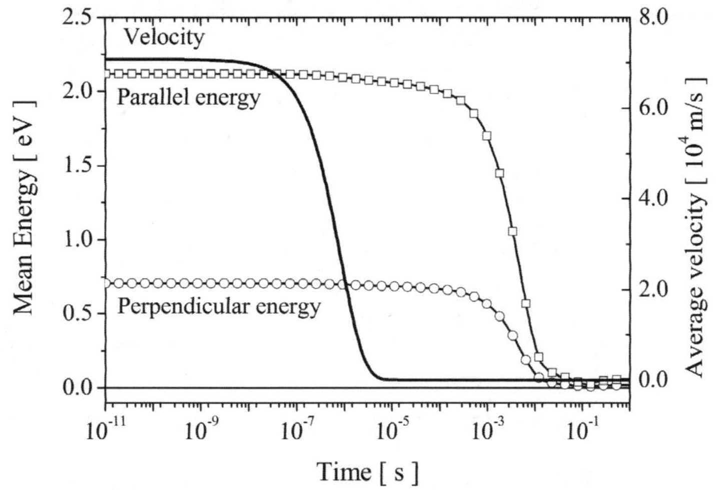

Figure 4. The relaxation times for thermalization of positrons in pure nitrogen.

3.2. Thermalization of the positrons

One of the key issues for optimum design and control of the positron traps is a knowledge of the relaxation times for thermalization of the positrons. As an initial study we investigated the relaxation time of positrons in pure nitrogen using a set of cross sections compiled in our laboratory. The subject of our study is a swarm of positrons initially with a drifted Maxwellian distribution (1.5×10^4 K with a drift of 7.0×10^4 m/s) entering the chamber. In Fig. 4 we demonstrate the thermalization of the mean energy (parallel refers to the initial drift direction) and the drift velocity. We can see that the thermalization time for the drift velocity is of the order of microseconds as compared with the thermalization times for energy being of the order of 10s of milliseconds.

3.3. Spatial relaxation of the electrons

In this section we demonstrate the ability of a magnetic field to control the spatial relaxation of the electrons. We extend the previous work on the idealised steady-state Townsend experiment [21] to include the explicit influence of the angle between the fields on spatial relaxation of the electrons when non-conservative collisions are operative.

In the case of an electric field only, previous work on the electron spatial relaxation revealed the complex nature of the relaxation process and associated basic mechanisms. The nature of the spatial relaxation profiles is dependent on the interplay between the power dissipated in elastic collisional processes, power dissipated in threshold collisional processes and the power deposited into the swarm by the field [21]. For certain gases, there exists a "window" of electric field strengths where the relaxation profiles are damped oscillatory in nature, and outside this electric field window the profiles are monotonic.

Before considering the effects of a magnetic field on spatial relaxation of the electrons we present one particularly interesting set of results associated with the spatial relaxation of the electrons in gas mixtures. In Fig. 5 we show the spatial relaxation of the mean energy in pure argon and mixtures of argon and CF_4 at E/n_0 of 15 Td. The cross sections for argon are displayed and detailed in [22] while the cross sections for CF_4 are given in [23]. We observe that the mean energy in pure argon exhibits a damped oscillatory relaxation along a decaying profile. However, by introducing a small amount of molecular admixture (e.g. CF_4) the oscillations are firstly suppressed and then entirely quenched in the limit of higher concentration of CF_4. By introducing a molecular admixture other collision processes, preferentially lower threshold vibrational excitations are introduced. These new collision processes lead to more efficient damping than elastic collisions, by virtue of larger and different energy loss mechanisms.

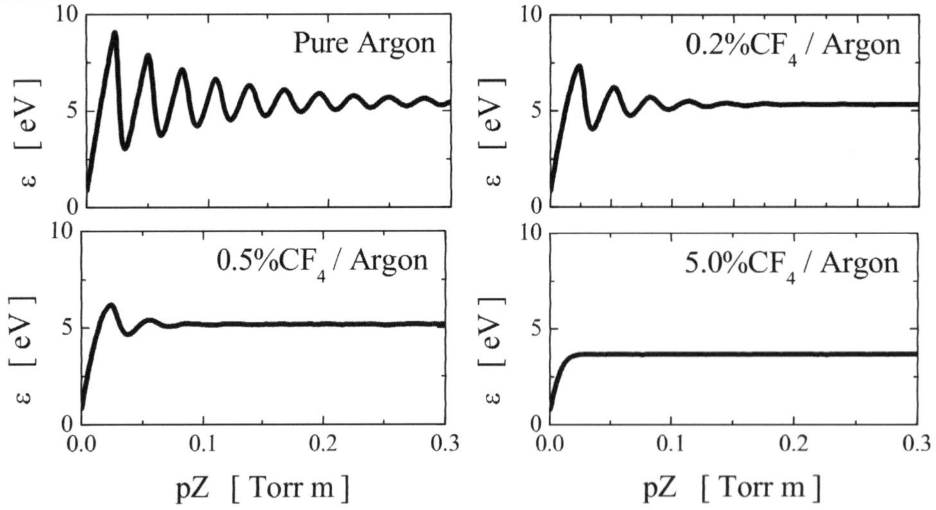

Figure 5. Spatial relaxation of the mean energy for pure argon and gas mixtures of argon and CF_4.

In Fig. 6 we show the effect of angle between electric and magnetic fields on spatial relaxation of the electrons for the Lucas-Saelee model. We observe that ε, v_z, v_x and v_y exhibit a damped oscillatory relaxation along a decaying profile for all ψ considered, although modulation is decreased with increasing ψ. We also note that v_x increases with the angle between the fields. This average velocity component has the worst statistic in our simulations. As ψ increases, the relaxation proceeds quicker while the period of oscillation becomes larger. As can be seen, the well known symmetry properties hold: $v_x = v_y = 0$ for $\varphi = 0^o$ and $v_y = 0$ for $\varphi = 90^o$. Another striking phenomenon is that v_y is comparable to v_x and v_z but yet it is common in the literature for plasma modellers to fail to include this quantity in their models. An extension of standard plasma models to include the (**E**×**B**)×**B** effects on electron transport may led to a better understanding of the power transfer to magnetically assisted/enhanced plasma reactors [24].

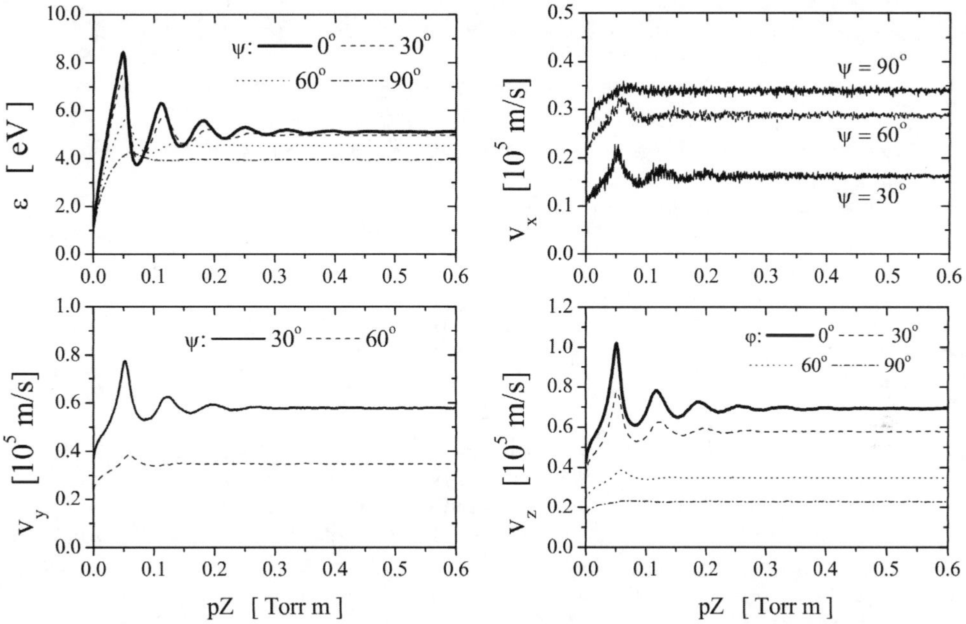

Figure 6. Spatial relaxation of the mean energy and average velocity components for Lucas-Saelee model at E/n_0 of 10 Td and B/n_0 of 200 Hx for various angles between the electric and magnetic fields as indicated at the graphs. The initial electron energy is 1.5 eV.

4. Concluding remarks

In this article we have presented a systematic study of the influence of magnetic fields and non-conservative collisional processes on the temporal and spatial relaxation properties of electron and positron swarms in gases. For the study of temporal relaxation under hydrodynamic conditions we have implemented (i) a time-dependent multi-term solution of Boltzmann's equation, and (ii) a time-resolved Monte-Carlo simulation. The results of the two independent techniques were compared for a benchmark gas and found to be in excellent agreement. In this study we focussed on (i) the explicit contribution of non-conservative processes to the bulk transport coefficients, and (ii) the influence of the angle between the electric and magnetic fields. The study yielded some interesting phenomena including the existence of transient negative diagonal elements of the diffusion tensor. In addition, we demonstrated that the traditional two-term approximation was invalid under certain conditions. As an interesting application of our theory, we also applied to study the relaxation time of positrons in molecular nitrogen. This is of particular relevance to understanding positron traps. For the study of

Journal of Physics: Conference Series **115** (2008) 012017 doi:10.1088/1742-6596/115/1/012017

spatial relaxation under non-hydrodynamic conditions we implemented a spatially-resolved Monte-Carlo simulation. We demonstrated that we were able to suppress the traditional Franck_Hertz oscillations observed in pure Ar by the introduction of a small fraction of CF_4 into the system. In addition, for the first time, we have presented results that investigate the effect of varying the angle between the electric magnetic fields on the spatial evolution of the swarm. It was found that the nature of the profiles could be controlled by varying the angle: Frank-Hertz oscillations could be suppressed while oscillation periods could be modified.

References

[1] Winkler R, Loffhagen D and Sigeneger F 2002 *Appl. Surf. Sci.* **192** 50
[2] Warman J M, Sowada U and De Haas M P 1985 *Phys. Rev. A* **31** 1974
[3] White R D, Dujko S, Ness K F, Raspopović Z M, Petrović Z Lj and Robson R E, to be published
[4] Nakano N, Shimura N, Petrović Z Lj and Makabe T 1994 *Phys. Rev E.* **49** 4455
[5] Petrović Z Lj, Raspopović Z M, Dujko S and Makabe T 2002 *Appl. Surf. Sci.* **192** 1
[6] White R D, Ness K F and Robson R E 2002 *Appl. Surf. Sci.* **192** 25
[7] Boeuf J P and Marode E 1982 *J. Phys. D: Appl. Phys.* **15** 2169
[8] Kim J B, Kawamura K, Choi Y W, Browden M D and Muraoka K 1999 IEEE *Trans. Plasma Sci.* **27** 1510
[9] Boltzmann L 1872 *Wein. Ber.* **66** 275
[10] Wang-Chang C S, Uhlenbeck G E and DeBoer J 1964 *Studies in Statistical Mechanics* vol 2, ed J. DeBoer and G.E. Uhlenbeck (New York: Wiley) p. 241
[11] Robson R E and Ness K F 1986 *Phys. Rev. A* **33** 2068
[12] White R D, Ness K F, Robson R E and Li B 1999 *Phys. Rev. E* **20** (2) 2231
[13] Sakai Y, Tagashira H and Sakamoto S 1977 *J. Phys. D: Appl. Phys.* **10** 1035
[14] Dujko S, White R D, Ness K F, Petrović Z Lj and Robson R E 2006 *J. Phys. D: Appl. Phys.* **39** 4788
[15] Loffhagen D and Winkler R 1999 IEEE *Trans. Plasma Sci.* **27** 1262
[16] White R D, Li B, Dujko S, Ness K F and Robson R E 2006 *The Physics of Ionized Gases*, Book of Invited Lectures, Topical Invited Lectures and Progress Reports, ed Lj. Hadžievski, B.P. Marinković and N.S. Simonović, 23rd Summer School and International Symposium on the Physics of Ionized Gases, (National Park Kopaonik, Serbia, 28 August-1 September 2006), (New York: American Institute of Physics) p.51
[17] Reid I D 1979 *Aust. J. Phys.* **32** 231
[18] Lucas J and Saelee H T 1975 *J. Phys. D: Appl. Phys.* **8** 640
[19] Raspopović Z M, Sakadžić S, Petrović Z Lj and Makabe T 2000 *J. Phys. D: Appl. Phys.* **33** 1168
[20] White R D, Ness K F and Robson R E 1999 *J. Phys. D: Appl. Phys.* **32** 1842
[21] Li B, Robson R E and White R D 2006 *Phys. Rev. E.* **74** 026405-1.
[22] Petrović Z Lj, Šuvakov M, Nikitović Ž, Dujko S, Šašić O, Jovanović J, Malović G and Stojanović V 2007 *Plasma Sources Sci. Technol.* **16** S1-S12.
[23] Kurihara M, Petrović Z Lj and Makabe T 2000 *J. Phys. D: Appl. Phys.* **33** 2146
[24] Dujko S, Petrović Z Lj, Raspopović Z M, White R D, Ness K F and Robson R E 2007, ed J. Schmidt, M. Simek, S. Pekarek and V. Prukner, Published by Institute of Plasma Physics AS CR v.v.i.; (2007) pp.265-268.

Journal of Physics: Conference Series **115** (2008) 012018 doi:10.1088/1742-6596/115/1/012018

Low energy electron attachment to condensed formic acid

Sylwia Ptasińska[1, 2]*, Andrew D. Bass[1], and Léon Sanche[1]

[1] Groupe en Sciences des Radiations, Département de Médecine Nucléaire et de Radiobiology, Faculté de Médecine, Université de Sherbrooke, Québec, Canada J1H 5N4
[2] Department of Physics and Astronomy, The Open University, Milton Keynes, United Kingdom MK7 6AA

Abstract

Dissociative electron attachment to formic acid in the condensed phase is studied using improved mass spectrometric detection of the negative ion fragments. The desorbed yields are measured as a function of incident electron energy in the range between 3 to 20 eV. Unlike previous work, the formation of the dehydrogenated anion $HCOO^-$ is observed and the signal to noise ratio is much higher for all other ions detected, i.e. OH^-, O^- and H^-. Resonant structure seen in all anion yield functions, is attributed to dissociative electron attachment (DEA), whereas above 14 eV nonresonant dipolar dissociation (DD) dominates the desorption yields.

*corresponding author: s.ptasinska@open.ac.uk

Introduction

Formic acid contains a commonly found structural feature with many complex molecules, namely the carboxylic group -COOH, it can thus serve as a model system. It is therefore of interest in many areas of physics and chemistry, more specifically in the fields of astrophysics and biochemistry. The molecule's presence of formic acid in interstellar clouds was discovered by means of micro-wave rotational transitions, in emission towards the galactic center source of Sagittarius B2 [1,2,3,4]. Traces of this simplest organic acid have also been detected in the coma of the Hale-Bopp comet [5,6] and in massive star forming regions [7,8,9,10,11]. In addition to these exciting occurrences in observational astronomy, formic acid plays a significant role in many biochemical processes and is a building block of various natural products. Thus, the interaction of low energy electrons (LEE) with condensed formic acid can be of particular interest in the understanding of radiation induced chemistry both in the interstellar medium and in biological systems. It has been already found that DNA damage by LEE, involves electron capture by a particular DNA components followed by the decay into dissociative electron attachment (DEA) and other dissociative channels [12]. Moreover, the common feature in the DEA cross sections of formic acid and some selected amino acids was attributed to the attachment into the –COOH π^* orbital, but very recently also the evidence of direct attachment to the lowest σ^* orbital was reported [13].

There are many recent experimental studies concerning DEA to the monomeric [14,15,16,17,18,19] and cluster [20] forms of formic acid in the gas phase. Those studies have shown the formation of negative fragment ions i.e. $HCOO^-$, OH^-, O^- and H^- in the electron energy range of 0-15 eV. Moreover, the formation and desorption of the H^-, O^-, OH^- ions from condensed HCOOH was also reported previously [21]. The ion yield of the H^- ion recorded as a function of the electron energy exhibited a broad resonant structure located around 9-10 eV. The similar ion yield functions for H^-, regarding to the shape and the position of maximum of the peak, have been observed from larger organic acids in their condensed phase, e.g. acetic acid (CH_3COOH) [22]. Measurements of the desorption of hydrogen/deuterium ions from isotopically labeled acetic acid molecules showed a more efficient production of H^-/D^- from the acid group –COOH(D) than from the methyl group - CH_3/ CD_3. Generally, the desorption of heavier mass anions is not observed due to their typically having insufficient kinetic energy to overcome the attractive polarization and image-charge forces induced in the molecular film close to the metal substrate. In the present study, by utilizing a modified system for the detection of ions not only ions desorbed with sufficient kinetic energy can be detected but also heavier ions formed on the surface of the multilayer film of formic acid.

Experimental setup

The present data were recorded with a new type of electron stimulated desorption apparatus which will be described in more details elsewhere [23]. The experiments were carried out in a stainless steel ultrahigh vacuum chamber, ion pumped to a base pressure of 5 x 10^{-10} Torr. Pure formic acid films are formed by condensation directly onto a clean polycrystalline Pt substrate cooled to 15 K by a closed cycle He compressor. The temperature is monitored by an Ag-Ni thermocouple attached to the cold tip of the cryostat. The Pt substrate consists of a 2.0 x 1.0 cm^2 foil of 0.0075 mm nominal thickness and having a purity of 99.95% (Goodfellow Cambridge Limited). The foil can be cleaned by resistive heating to 900 K. A gas-handling manifold consists of two gas or vapor sources connected through bypass and precision leak valves to a small volume. The absolute pressure in this volume is measured with a barometer. Known amounts of gas or vapor are leaked into the analysis chamber by monitoring the differential pressure drop in this volume. The latter is connected via an admission valve to stainless steel capillary (a 2 mm inside diameter) which ends approximately 15 mm from the surface of the Pt. The details of the volumetric dosing procedure have been described previously [24]. A film thickness,

Journal of Physics: Conference Series **115** (2008) 012018 doi:10.1088/1742-6596/115/1/012018

determined by this method, has an uncertainty of about ± 30% with overall repeatability of ± 0.2 monolayer (ML). Formic acid was obtained from Sigma-Aldrich with a stated purity of 98 %. The acid was additionally degassed by repeated freeze-thaw cycles carried out while pumping.

The sample films were bombarded with a 3-20 eV pulsed electron beam of about 3 nA time-averaged current that was produced by a Kimball Physics ELG-2 electron gun. Electrons impinged onto the sample in the horizontal plane at an incident angle of 45° with respect to the surface normal and were focused into a 1 mm^2 spot from a 20 mm working distance (as measured at an incident energy of 10 eV). The absolute electron energy scale of the incident beam was calibrated by taking 0 eV as the onset of the current transmitted at the platinum surface, with an estimated error of about ± 0.3 eV. The electron gun with pulse capability (rise/fall time ~ 10 ns, pulse width ~ 20 ns to 100 μs) is synchronized with the sweep counter of a time of flight (ToF) mass analyzer so that mass spectra can be measured at different incident electron energies. The anion detection is achieved by applying a negative potential pulse (-2 kV, rise/fall time 30 ns, pulse width of 2 μs) applied to the Pt substrate within a short time (~ 10 ns) after the end of each pulse of incident electrons. Desorbed ions are propelled into the entrance optics of a Reflectron ToF mass analyzer (Kore Technology R-500) positioned along the surface normal, at 10 mm from the sample surface. Such an experimental set up allows ions to be detected with high sensitivity, while inflicting a minimum amount of damage to the condensed molecular film.

Results

LEE can initiate molecular fragmentation by either ionization, excitation, or electron attachment to the molecules. It is generally accepted that the basic mechanisms responsible for the formation and desorption of negative ions following low energy electron impact on thin molecular films can be related to elementary process known from gas phase studies, namely dissociative electron attachment (DEA) and dipolar dissociation (DD) [25]. DEA is initiated by the resonant attachment of an electron to a molecule which leads to the formation of a transient negative ion $(TNI)^{\#-}$. The latter may either stabilize, autodetach the electron, or decay into negative ion and neutral (radical) fragments.

In the present experiments the following DEA reaction channels would seem to operate in the electron energy range from 3 to 20 eV:

$$e^- + HCOOH \rightarrow (HCOOH)^{\#-} \rightarrow H^- + HCOO \quad (1a)$$
$$\rightarrow O^- + H_2CO \quad (1b)$$
$$\rightarrow OH^- + HCO \quad (1c)$$
$$\rightarrow HCOO^- + H \quad (1d)$$

Ion yields as a function of incident energy for all desorbed ions are presented in Figs 1-4, respectively, and the positions of all resonances observed are listed in Table 1.

	Peak positions (eV)		
H^-	9.5		
O^-	10.0	13.5	17.5
OH^-	11.3	17.5	
$HCOO^-$	12.7	17.2	

Table 1. Electron energy position (in eV) of resonances for all observed fragment anions formed via DEA to formic acid.

The strongest signal is that of H^- desorption represented by channel (1a). The ion yield function in Fig. 1 shows an intense H^- signal peaking at 9.5 eV, similar to that observed in the previous measurement [21], whereas, two features were observed, a main peak at 7.3 and a shoulder at about 9 eV in the H^- yield in the gas phase [18]. The formation of the H^- ion can be attributed to a core excited resonance, which involves an electronically excited species with two electrons in usually unoccupied molecular orbitals. There are two possible sites in the formic acid molecule from which H^- can be desorbed via dissociation of O-H or of C-H bonds. Both channels are also thermodynamically available at energy thresholds of 3.79 and 3.43 eV, respectively [18]. In the present experiment the appearance energy for H^- is found to be about 5 eV, within the limit of the energy resolution. Despite the fact that experiments were undertaken with two isotopically labeled molecules, i.e. HCOOD and DCOOH, no particular selectivity has been observed, mainly due to strong hydrogen/deuterium scrambling prior to dissociation [21].

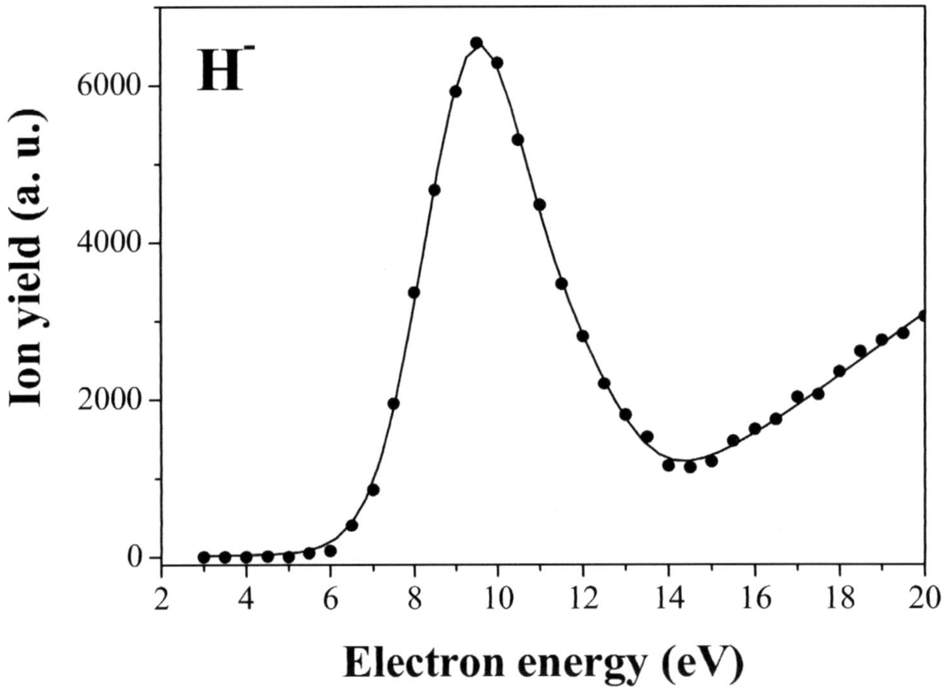

Fig. 1. The hydrogen (H^-) ion yield function from a formic acid film measured in the electron energy range between 3 to 20 eV.

The formation of the O^- ion via reaction (1b) is also thermodynamically possible with a threshold of 3.84 eV [14]; however, the signal detected for that ion has been observed as very weak from HCOOH in both the gas phase [14,18] and the condensed phase [21]. The ion yield detected in the latter case showed two features located at about 10 and 17 eV, similar to those observed here (Fig. 2). However, the previous study [22] also showed that this O^- signal evolved during irradiation and that a O^- signal could not be detected from a freshly deposited sample of HCOOH. Therefore, it was concluded that the

O^- signal derived from DEA to an irradiation product, likely CO_2 [21]. Due to this fact, O^- desorption will not be further discussed.

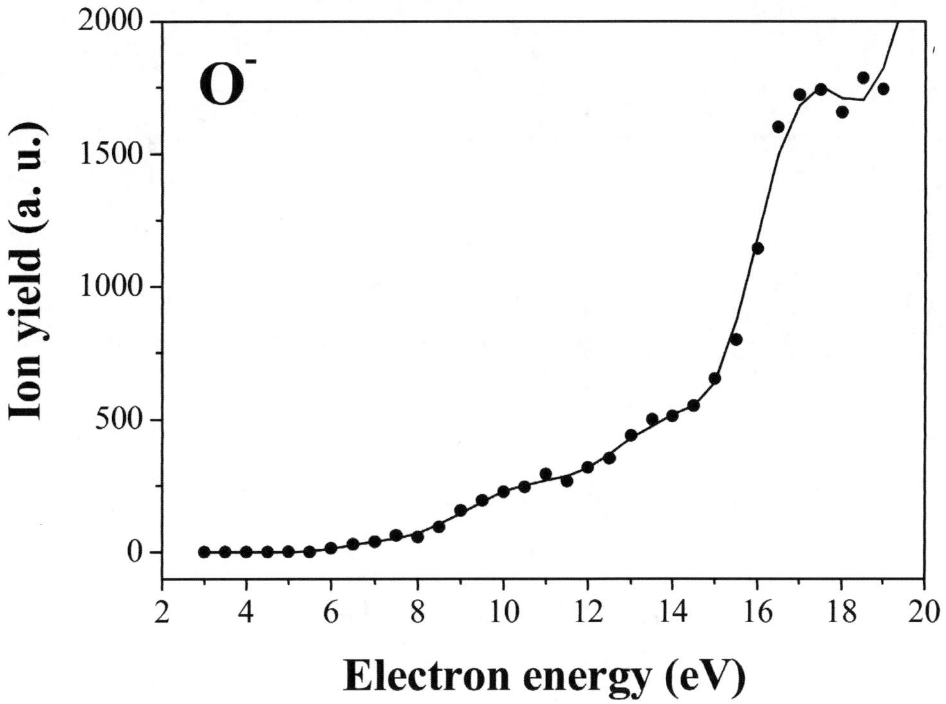

Fig. 2. The oxygen (O^-) ion yield function from a formic acid film measured in the electron energy range between 3 to 20 eV.

Fig. 3 presents the OH^- ion desorption yield as a function of incident energy obtained from the formic acid film. The strongest OH^- desorption signal is obsrved between 9 and 13 eV, where a broad resonant structure is observed, as in the previous experiment [21], but with a much larger signal-to-noise ratio. The gas phase studies [14,18] also showed the formation of the hydroxyl anion for HCOOH, but at much lower energy, i.e. around 7.5 eV with an estimated thermodynamic threshold of 3.48 eV and the simultaneous generation of the HCO radical. In all cases presented in Figs. 1-4, the anion signal observed in the yield function above 15 eV is principally due to the DD process.

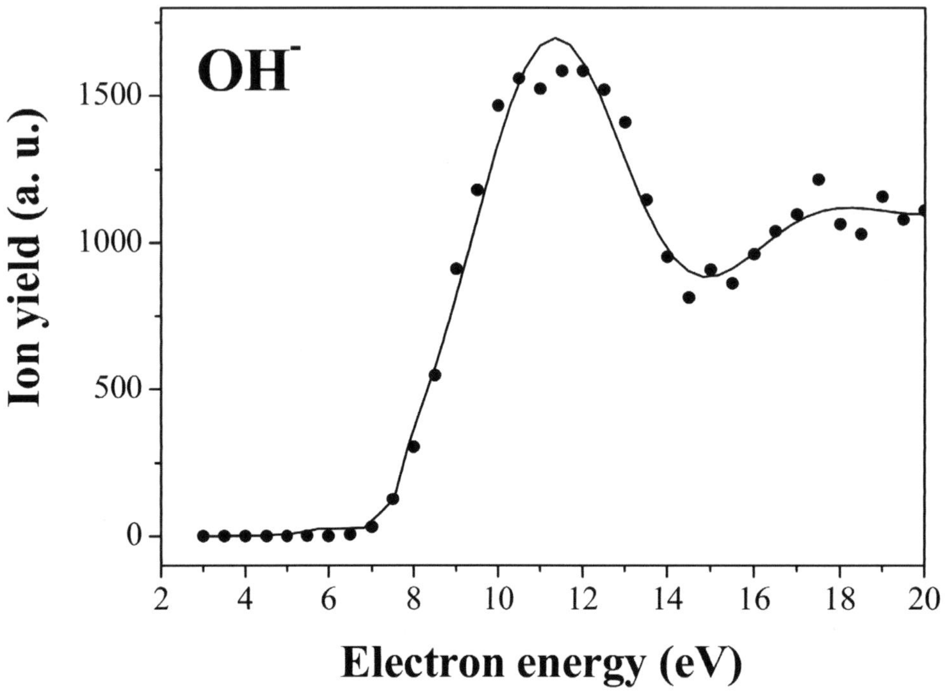

Fig. 3. The hydroxyl (OH^-) ion yield function from a formic acid film measured in the electron energy range between 3 to 20 eV.

Previous studies on anion desorption from formic acid has not shown any $HCOO^-$ signal, however authors have noticed that the reaction channel (1d) could be possible also in the condensed phase, but probably is suppressed by insufficient kinetic energy of that fragment to overcome the polarization force [21]. In the present work, due to the higher sensitivity, the signal of the dehydrogenated anion of formic acid is detected as a function of electron energy (Fig. 4). The $HCOO^-$ ion yield shows two resonant structures near 12.5 and 17.5 eV. In contrast to this high energy process observed in the condensed phase, the formyloxyl anion ($HCOO^-$) formed via DEA to formic acid in the gas phase hasits highest yields at low energy with a resonance maximum located at 1.25 eV [14,18,20]. Due to the considerable electron affinity of the HCOO radical (3.5 eV) electron capture at 1.25 eV forming the dehydrogenated anion occurs via a shape resonance, with the extra electron occupying the lowest orbital of π^* character and localized on the COOH group [20]. Gas phase studies on DEA to different isotopically labeled molecules of formic acids (i.e. HCOOD and DCOOH) showed a dehydrogenation reactions, where abstraction of neutral H or D is exclusively observed from O-H and O-D site, respectively [20].

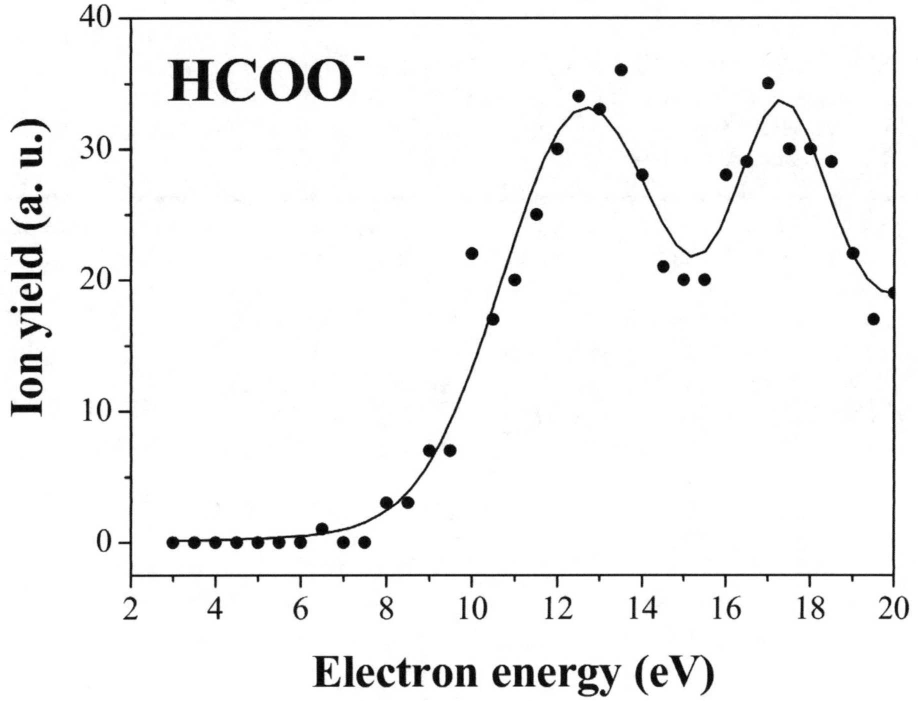

Fig. 4. The formyloxyl ($HCOO^-$) ion yield function from a formic acid film measured in the electron energy range between 3 to 20 eV.

In conclusion, we have presented the results of an investigation on dissociative electron attachment to condensed phase formic acid performed with a new apparatus having a higher sensitivity for condensed phase studies of electron-induced ion desorption than previously available. This advance in the technique allows more different types of desorbed anions to be detected with a higher signal to noise ratio [23]. As an example, we have shown that the present system could easily record the yield function of the dehydrogenated ion $HCOO^-$ desorbing from condensed formic acid, which was previously undetected in similar experiments. The yields of $HCOO^-$ and all other fragment anions recorded as a function of the incident energy in the range between 3 to 20 eV are due to the core excited resonance dominated DEA and DD processes. For the first time, the formation of the dehydrogenated ion $HCOO^-$ is observed in the condensed formic acid. Since formic acid is the simplest model biosystem and due to electron controlled functionalization this study may be considered to be important to understand electron induced damage to more complex biomolecules.

Acknowledgement

This work is financed by the Canadian Institutes of Health Research (CIHR). S.P. gratefully acknowledges financial support from CIHR and EPSRC in the form of a Postdoctoral Fellowship at the Sherbrooke University and The Open University, respectively.

References

1. B. Zuckerman, J. A. Ball, C. A. Gottlieb, Astrphys. J. 163 (1971) L41
2. G. Winnewisser, E. Churchwell, Astrophys. J. 200 (1975) L33
3. J. Ellder, P. Friberg, A. Hjalmarson, B. Hoeglund, W. M. Irvine, L. E. B. Johansson, H. Olofsson, G. Rydbeck, O. E. H. Rydbeck, Astrophys. J. 242 (1980) L93
4. W. M. Irvine, P. Friberg, N. Kaifu, Y. Kitamura, K. Kawaguchi, Astrophys. J. 342 (1989) 871
5. D. Bockelee-Morvan, D.C. Lis, J. E. Wink, D. Despois, J. Crovisier, R. Bachiller, D. J. Benford, N. Biver, P. Colom, J.K. Davies, E. Gerard, B. Germain, M. Houde, D. Mehringer, R. Moreno, G. Paubert, T. G. Phillips, H. Rauer, Astron. Astrophys. 353 (2000) 1101
6. S. D. Rodgers, S. Charnley, Monthly Notices R. Astron. Soc. 320 (2001) 1101
7. G. Winnewisser, E. Churchwell, Astrophys. J. Lett. 200 (1975) L33
8. S.-Y. Liu, J. M. Girart, A. Ramijan, L. E. Synder, Astrophys. J. 576 (2002) 255
9. S.-Y. Liu, D. M. Mehringer, L. E. Snyder, Astrophys. J 552 (2001) 654
10. J. M. Hollis, J. A. Pedelty, L. E. Synder, P. R. Jewell, F. J. Lovas, P. Palmer, S.-Y. Liu, Astrophys. J. 588 (2003) 353
11. S. Cazaux, A. G. G. M. Tielens, C. Ceccarelli, A. Castets, V. Wakelam, E. Caux. B. Parise, D. Teyssier, Astrophys. J. Lett. 593 (2003) L51
12. B. Boudaiffa, P. Cloutier, D. Hunting, M. A. Huels, L. Sanche, Science 287 (2000) 1658
13. A. M. Scheer, P. Mozejko, G. A. Gallup, P. D. Burrow, J. Chem. Phys. 126 (2007) 174301
14. A. Pelc, W. Sailer, P. Scheier, M. Probst, N. J. Mason, E. Illenberger, T. D. Märk, Chem. Phys. Lett. 361 (2002) 277
15. A. Pelc, W. Sailer, P. Scheier, N. J. Mason, E. Illenberger, T. D. Märk, Euro. Phys. J. D 20 (2002) 441
16. A. Pelc, W. Sailer, P. Scheier, N. J. Mason, E. Illenberger, T. D. Märk, Vacuum 70 (2003) 429
17. A. Pelc, W. Sailer, P. Scheier, T. D. Märk, Vacuum 78 (2005) 631
18. V. S. Prabhudesai, D. D. Nandi, A. H. Kelkar, R. Parajuli, E. Krishnakumar, Chem. Phys. Lett. 405 (2005) 172
19. V. S. Prabhudesai, A. H. Kelkar, D. D. Nandi, E. Krishnakumar, Phys. Rev. Lett. 95 (2005) 143202
20. I. Martin, T. Skalicky, J. Langer, H. Abdoul-Carime, G. Karwasz, E. Illenberger, M. Stano, S. Matejcik, Phys. Chem. Chem. Phys. 7 (2005) 2212
21. T. Sedlacko, R. Balog, A. Laffose, M. Stano, S. Matejcik, R. Azria, E. Illenberger, Phys. Chem. Chem. Phys. 7 (2005) 1277
22. M. Bertin, D. Cáceres, M. P. Davis, R. Balog, A. Lafosse, N. J. Mason, E. Illenberger, R. Azria, Chem. Phys. Lett. 433 (2007) 292
23. M. Bazin, S. Ptasinska, A. D. Bass, L. Sanche, in preparation
24. L. Sanche, J. Chem. Phys. 71, 4860 (1979)
25. Electron-Molecule Interactions and Their Applications, edited by L.G. Christophorou (Academic, Orlando, 1984) vol. 1

Journal of Physics: Conference Series **115** (2008) 012019 doi:10.1088/1742-6596/115/1/012019

Electron attachment to SF_6 under well defined conditions: Comparison of statistical modeling results to experiments

Thomas M. Miller[1], Albert A. Viggiano[1], and Jürgen Troe[2]

[1] *Air Force Research Laboratory, Space Vehicles Directorate, Hanscom Air Force Base, Bedford, Massachusetts 01731-3010, USA*

[2] Institute for Physical Chemistry, University of Göttingen, Tammannstrasse 6, D-37077 Göttingen, Germany

E-mail: miller.hanscom@gmail.com

Abstract. Experiments were carried out using a flowing-afterglow Langmuir-probe apparatus to measure rate constants for electron attachment to SF_6 and thermal detachment from SF_6^-. In a recent series of papers, these results were combined with new and existing data on nondissociative and dissociative attachment to SF_6 and compared to statistical modeling of the various processes involved in the stabilization of the ionic products of attachment. This paper gives a summary of those findings. The major conclusions are: (a) only the ground electronic state of SF_6^- needs to be invoked to explain available data; (b) the electron affinity of SF_6 is higher than previously thought, namely, EA(SF_6) = 1.20 (± 0.05) eV; (c) the endothermicity of the dissociative electron attachment reaction that yields SF_5^- is 0.41 eV (± 0.05) eV at 0 K; (d) combining these two numbers gives the bond energy $D_0^o(F—SF_5^-)$ = 1.61 (± 0.05) eV.

I. Introduction

In three earlier papers we delineated various aspects of electron attachment to SF_6 and the reverse detachment from SF_6^- under thermal and low-energy conditions (electron energies less than about 0.4 eV).[1-3] There exists also a wealth of information on the higher energy processes, but those are outside the scope of the present study. For a more detailed discussion of the higher-energy processes, see the reviews by Christophorou and Olthoff.[4,5] Our goal in the earlier papers was to present a complete and consistent picture of the numerous steps involved in the electron attachment reaction. This required new experimental and theoretical techniques. The flowing-afterglow Langmuir-probe apparatus (FALP) was used in new ways to make measurements on electron attachment and thermal detachment under truly thermal conditions over extended temperature and pressure ranges. In order to explain both the new and previous data, statistical theories that were developed for neutral and ion- molecule reactions were modified for use in understanding electron processes. The new measurements combined with the detailed modeling allowed derivation of the electron affinity of SF_6 and the bond energy $D_0^o(F—SF_5^-)$.

While the SF_6 system has been and continues to be the most studied electron attachment reaction, many aspects of the reactivity have appeared in conflict or remain unexplained. Due to the complexity of the system, the modeling presented in our earlier papers focused on the intricate detail needed to understand a wide range of experimental observations. In this paper, we will present the current understanding of the electron attachment reaction with SF_6, passing over the details of the measurements and modeling. We are only focused on thermal or low-energy processes for both associative and dissociative attachment to SF_6 as well as thermal detachment from SF_6^-. The regime of interest is sometimes referred to as the zero-energy resonance. Higher energy resonances include formation of SF_4^-, SF_3^-, SF_2^-, F_2^-, and F^- and are not discussed. In addition, we focus on the most informative results and do not intend to give a comprehensive review—see Christophorou and Olthoff.[4,5]

Our earlier series of papers clearly showed the need for a complete characterization of the experimental conditions used in making measurements on this reaction, including the average electron energy, as well as the electron energy distribution, the temperature of SF_6, the number density and identity of the gas in the interaction region, and the transit time from the interaction region to the detector. Thermal measurements always have the conditions well defined, though not necessarily specified in publications. In contrast, beam experiments can be difficult to interpret unless all parameters are not only well defined but accounted for in the data analysis. In order to present an overall, consistent picture of the reaction, we will discuss a number of pitfalls that can occur in measurements.

II. Production of SF_6^-

The overall reaction scheme is shown in figure 1. The first step in the reaction sequence is electron capture into a resonant continuum state, which we denote by e^-SF_6,

$$e^- + SF_6 \rightarrow e^-SF_6 \qquad \textit{capture.} \tag{1}$$

The description of step (1) for SF_6 is provided by pure *s*-wave capture, which is commonly modeled by the formula due to Klots[6] which deviates by as much as 8% from the correct Vogt-Wannier *s*-wave cross section.[7] A new, more accurate formula for the *s*-wave capture cross section $\sigma_c(E)$ as a function of electron energy E has been presented by Dashevskaya et al.,[8]

$$\sigma_c(E) = (\pi/2E)[1 - 0.5\exp(-\sqrt{8\alpha E}) - 0.5\exp(-\sqrt{72\alpha E})] \qquad \textit{s-wave capture,} \tag{2}$$

where α is the molecular average polarizability in a_o^3, E is the electron energy in hartree, and $\sigma_c(E)$ is in a_o^2. The new formula has deviations of <2% from the Vogt-Wannier *s*-wave capture cross section. Dashevskaya et al. also presented formulas for capture involving higher-order partial waves and ion-dipole interactions (not needed for low-energy attachment to SF_6).[8]

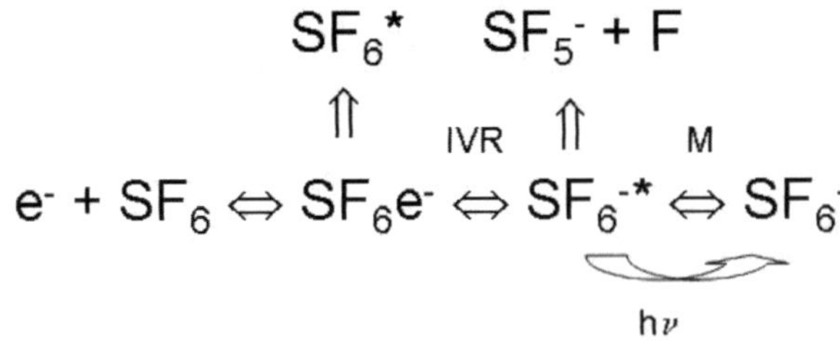

Figure 1. Scheme showing the individual processes involved in the electron-SF_6 reaction.

The nascent e^-SF_6 can then undergo intramolecular vibrational redistribution (IVR) to form highly vibrationally excited SF_6^{-*} in its ground electronic state, with the total energy still in the continuum,

$$e^-SF_6 \rightarrow SF_6^{-}* \qquad \textit{IVR.} \qquad (3)$$

The reverse of process (3), followed by the reverse of process (1), is detachment

$$SF_6^{-*} \rightarrow e^-SF_6 \rightarrow e^- + SF_6 \qquad \textit{detachment.} \qquad (4)$$

Additionally, electrons can vibrationally excite the neutral SF_6 (shown as SF_6* in figure 1)[7,9].

$$e^- + SF_6 \rightarrow e^-SF_6 \rightarrow SF_6* + e \quad \textit{vibrational excitation.} \qquad (5)$$

In order to form stabilized SF_6^-, i.e., with a total energy below the continuum, energy needs to be dissipated from SF_6^-* through radiation

$$SF_6^-* \rightarrow SF_6^- + h\nu \qquad \textit{radiative stabilization} \qquad (6)$$

or through collisions with a third body,

$$SF_6^-* + M \rightarrow SF_6^- + M \qquad \textit{collisional stabilization.} \qquad (7)$$

In addition, a dissociative channel, yielding SF_5^- + F, can open up. That process will be discussed in the next section.

Each process listed above is important in determining cross sections and rate constants for SF_6^- production in some pressure, temperature, or energy regime. In our earlier papers, the modeling of each step was presented, either through statistical theory or from calibration to experimental data.[1-3] In particular, detachment, IVR, and vibrational excitation cross sections are difficult to calculate from first principles, so the relevant parameters were determined from measured cross sections or rate constants. Likewise, the radiative contribution, equation (6), was calibrated using ion-cyclotron resonance (ICR) measurements of the stabilization of SF_6^-* at very low pressures. A master equation governing energy gain and loss in collisions was used to model collisional stabilization [equation (7)], with parameters determined from ion-molecule reaction studies.

The first set of experimental data discussed here are rate constants as a function of temperature for the overall attachment process in the high pressure limit. There is general agreement on the 300 K rate constants with many groups finding values near the very accurate measurements of Crompton and coworkers,[10,11] and for this reason, relative cross section measurements taken as a function of electron energy are usually normalized to the Crompton value of $2.27 \pm 0.07 \times 10^{-7}$ cm^3 s^{-1} at 295 K. Crompton and coworkers also measured a rate constant of $2.20 \pm 0.09 \times 10^{-7}$ cm^3 s^{-1} at 500 K, indicating that there is little temperature dependence, and imperceptible difference between buffer gases, in their pressure range of 8-23 Torr. However, low temperature measurements carried out with the CRESU (Cinétique de Réaction en Ecoulement Supersonique Uniforme) technique[12] showed an unexpected temperature dependence. Figure 2 shows the complete temperature range of available data along with the s-wave capture curve. The rate constants from 48-173 K lie well below the canonical room temperature value. In particular, from 48 to 123 K, rate constants were found to be independent of temperature at about 60% of the Crompton et al. values. From 123 to 300 K, the measured rate constants increased. It should be noted that T_{el} in the CRESU experiment was later found to range from T_{gas} up as high as 500 K due to superelastic collisions between electrons and vibrationally-excited N_2 buffer gas.[13] This contamination is not thought to affect CRESU results for SF_6 for T_{gas} in the 48-123 K range, at least, because the measured attachment rate constants appear independent of T_{gas}. New results in the temperature range 300-670 K from our laboratory were presented in our earlier series of papers[1,3], and show good agreement with Crompton et al. in the overlapping region and begin to fall slightly above 550 K, to a value of 2.0×10^{-7} cm^3 s^{-1} at 670 K. Many older measurements are cataloged in the reviews by Christophorou and Olthoff,[4,5] which show that the attachment rate constant falls continuously for temperatures above 700 K. The temperature data along with electron temperature data, and modeling results are plotted in Figure 3 as a function of electron temperature. The modeling results include all steps outlined in equations 1, 3-7. The model well represents the data for electron attachment to SF_6 in the high-pressure limit.

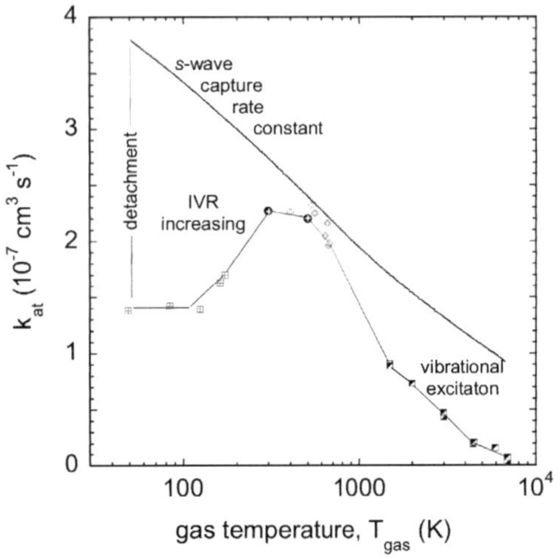

Figure 2. Total rate constants for electron attachment to SF_6, in the high-pressure limit, versus gas temperature. The departure from the *s*-wave capture rate constant provides input for modeling, as indicated (roughly) by the labeling. The data from 48-300 K are from Ref. 12. Those from 300-670 K are the present FALP data, except for the large dark circles (Refs. 10 and 11). Data at high electron temperature from ORNL are used to indicate trends expected for high T_{gas} (Refs. 4 and 5).

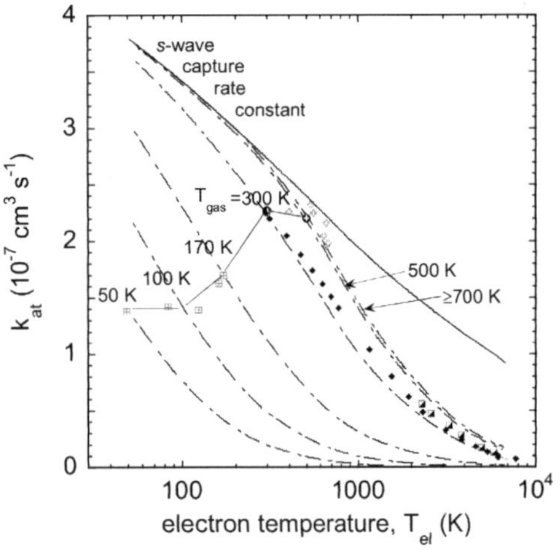

Figure 3. Predicted total rate constants for electron attachment to SF_6, in the high-pressure limit, versus electron temperature, for the fixed gas temperatures indicated by labels. The data shown are the same as in Fig. 2 except for those showing the T_{el} dependence (♦ for T_{gas} = 300 K as recommended in Refs. 4 and 5, and ◘ for T_{gas} = 500 K from Ref. 13).

Modeling the thermal attachment rate constants by averaging the Vogt-Wannier cross sections of equation (2) over a Boltzmann distribution of electron energies E leads to larger values than measured. This discrepancy at low temperatures is attributed to incomplete IVR where this process competes with autodetachment. Initially, the IVR reduction factor was modeled by a value that was independent of gas temperature. This proved inadequate to fit the temperature dependence shown in figure 2. Instead, it was necessary to have the IVR rate increase with increasing temperature, i.e., dependent upon the degree of vibrational excitation of the SF_6 molecule. The *s*-wave rate constant decreases monotonically with temperature, but the attachment rate constant increases mainly because the IVR efficiency increases. The rate constant eventually comes close to the *s*-wave collisional value, then decreases faster than the collisional rate constant with temperature, as more electron collisions go to vibrational excitation without attachment. The experimentally fitted IVR probability, with coefficients varying with temperature, were given in equation (3.6) and Table 3 of Ref. 1. At high temperatures, the discrepancy between modeled capture rate constants and experimental data is attributed to the competition between attachment and vibrational excitation, equation (5). Tentatively, the corresponding reduction factor was taken independent of temperature.

The modeling of the various processes versus electron energy allowed us to predict how the attachment rate constants would change for electron temperature (T_{el}) different from the gas temperature (T_{gas}). These results are shown in figure 3 along with data from the Oak Ridge National Laboratory (ORNL) recommended in Ref. 4, for T_{gas} = 300 K, and ORNL data from Ref. 14 for T_{gas} = 500 K.

III. Pressure Dependence

The discussion immediately above dealt exclusively with the high pressure limit of the reaction. A complete understanding, and therefore a faithful modeling of the rate constants, must also include pressure dependences. Much of the literature acknowledges that stabilization must be included in the attachment mechanism but then ignores the stabilization step in the interpretation of results, based on the general agreement of experimental data obtained in buffer gas pressures from 1 Torr to several atm. In the absence of radiative stabilization, the attachment rate constants should increase linearly with pressure (low pressure limit), then increase at a rate slower than pressure (falloff region), and finally saturate so that further increases in pressure do nothing (high pressure limit). Radiative stabilization is evident in the ICR data of Foster and Beauchamp.[15] At 300 K, they found that the attachment rate constant was 1.6 (± 0.2) × 10^{-8} cm^3 s^{-1}, independent of SF_6 pressure below 5 × 10^{-7} Torr. The long observation time (20-80 ms) in the ICR enabled those measurements to be made. The low pressure rate constant is only 7% of the value measured at high pressures. As N_2 gas was added to the cell, the attachment rate constant increased as expected, reaching 4.8 × 10^{-8} cm^3 s^{-1} at 7.3 × 10^{-5} Torr N_2.

The pressure dependence can be represented using the master equation approach of statistical theory, as formulated in Ref. 1 for the SF_6 problem. The master equation describes the effect of collisions on SF_6^-*, in competition with autodetachment, radiative stabilization, and dissociation (into SF_5^- + F). Collisions with buffer gas may increase or decrease the internal energy of SF_6^-*, but there is a net decrease, on average. The modeling requires knowledge of this average decrease in internal energy. A value of 200 cm^{-1} for a N_2 buffer gas was used, based on modeling of unimolecular decomposition of vibrationally excited ions.[16] Expressions for the pressure dependence were given as equations (6.1) and (6.2) in Ref. 1. The resulting falloff curves are shown in figure 4 (dashed lines) for an N_2 buffer gas at a variety of temperatures. The effect of radiative stabilization is shown by the solid curves in figure 4, based on modeling of the Foster and Beauchamp[15] ICR data at 300 K, yielding a radiative rate constant of 53 s^{-1}. An analysis based on *ab initio* calculations of radiative rates for vibrational states gave a value of 80 s^{-1},[17] in very good agreement with the modeled value. The model results are given in figure 4 and show that at low temperatures attachment rate constants approach the high pressure limit already at buffer gas concentrations corresponding to mTorr pressures. By 700 K, nearly 1 atm of N_2 is required to place an experiment in the high pressure limit. For a 300 K target gas, the radiative limit is reached only at concentrations below about 10^{-6} Torr.

Literature data show no apparent influence on the type of buffer gas used in experiments,[4,11] which is understandable in that almost all data to date have been obtained in or near the high pressure limit. If additional ICR data could be obtained at μTorr pressures, it is expected that differences could be observed, because the average energy $\langle E \rangle$ transferred per collision in the master equation describing collisional relaxation will be different for different gases. Beam data in the lowest energy regime simulate the high pressure limit since the lifetime for autodetachment is greater than the observation window. As the energy increases this is no longer true, and the data not only reflect the fundamental process being studied but also the instrument time window.

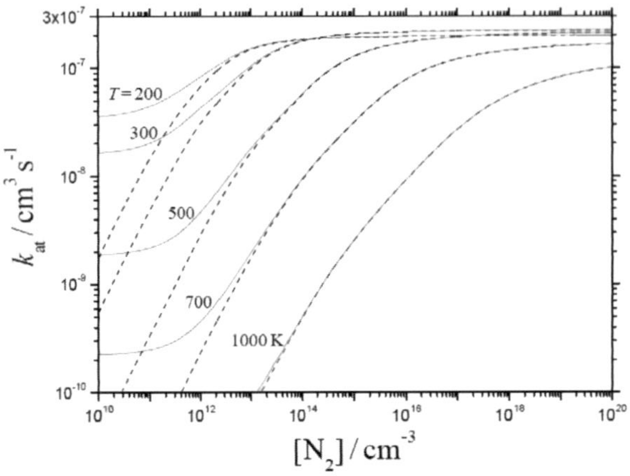

Figure 4. Predicted pressure dependence of rate constants for electron attachment to SF_6 for various gas temperatures. The solid curves include radiative stabilization of the SF_6^{-*}

IV. SF_5^- Production

SF_5^- may be produced following the initial attachment event in a slightly endothermic process.

$$SF_6^{-*} \rightarrow SF_5^- + F \qquad \textit{dissociation.} \qquad (8)$$

The SF_5^- branching fraction must be dependent on pressure since there is competition between SF_6^{-*} undergoing detachment, collisional and radiative stabilization, and dissociation. Previously, only the Fehsenfeld,[18] Foster and Beauchamp,[15] Crompton and Haddad,[10] and Petrović and Crompton[11] studies mentioned looking for a pressure dependence, aside from work conducted at very high pressures.[4,5] The pressure range in the Fehsenfeld experiment was 0.2-1.5 Torr (He gas) and 0.1-0.4 Torr (Ar gas) and no significant change was seen in either the rate constants or branching fraction.[18] The Australian experiments were carried out over a pressure range of 8-23 Torr of N_2, CO_2, H_2, and He.[10,11] Measurements at ORNL (see Christophorou and Olthoff[4,5]) have been carried out between 1-16 atm of N_2, Ar, and Xe, and earlier drift tube work in that laboratory utilized a wide variety of molecular gases at atm pressure and higher. As discussed above, the low pressure ICR study found the overall rate constant to rise with pressure. When one combines the Fehsenfeld[18] and ICR[15,19] values for the SF_5^- fraction, one sees a distinct pressure dependence. At essentially zero pressure in an ICR study, 5% SF_5^- is observed but at 1 Torr only ~0.01% is observed.

Španěl et al.[20] in 1995 noted the large difference in the SF_5^- branching fraction between Fehsenfeld's experiment at 1 Torr and the very low pressure experiment of Chen and Chantry.[21] They showed that the difference was due to collisional stabilization at 1 Torr gas pressure. From an analysis using RRK theory (see Refs. 20 and 22), they derived an endothermicity of 0.12 eV for the production of SF_5^-. The statistical modeling of the present work yields a larger value for the endothermicity. The more advanced nature of the recent statistical modeling yields a bond strength of $D_0^o(F\text{-}SF_5^-) = 1.61$ (± 0.05) eV.[2] The latter figure, combined with an improved electron affinity, $EA(SF_6) = 1.20$ (± 0.05) eV (Sect. VI) gives an endothermicity of 0.41 eV (at 298 K) for appearance of the SF_5^- product in the electron attachment reaction.

Modeling results for the specific rate constants for autodetachment [equation (3)] and dissociation [equation (8)] are shown in figure 5. The curves start from the respective thresholds and increase dramatically with energy. Once energetically possible, dissociation quickly dominates, i.e., at energies <0.1 eV above threshold. To a large extent the vastly different slopes are a result of the fact that detachment is an s-wave-only phenomenon and therefore does not have multiple J states involved. Dissociation includes all partial waves.

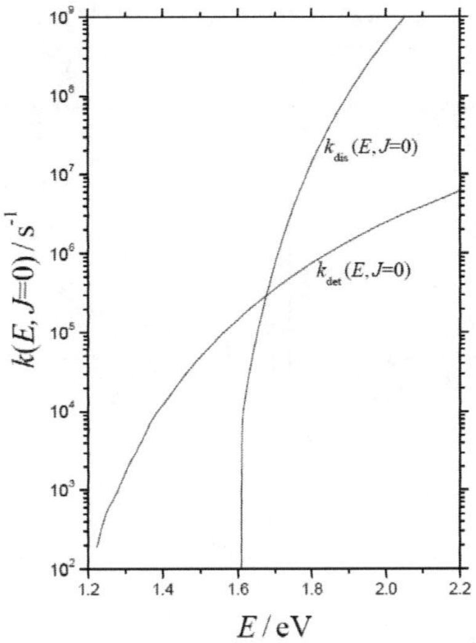

Figure 5. Specific rate constants for electron detachment of SF_6^{-*}, $k_{det}(E,J=0)$, and for dissociation to SF_5^- + F, $k_{dis}(E,J=0)$.

While the pressure dependence is rarely studied, the temperature and electron energy dependencies for SF_5^- production have long been recognized. In fact, the first mention of increase in SF_5^- fraction with higher temperature came already in 1958.[23] The temperature dependence is strikingly large. However, in order to compare the amount of SF_5^- formed in various experiments, it is important to look not only to the temperature and energy in an experiment, but also whether the SF_6^{-*} has had time to dissociate. The dissociation process will go to completion only in the presence of sufficient three-body collisions (to complete the stabilization of those SF_6^{-*} which haven't dissociated) or with a long observation window such as found in the ICR experiments. A long time window would allow radiative stabilization to come to an end and is therefore in the 0.1 s or longer range (several half-lives). Therefore, determining meaningful SF_5^- branching fractions in beam experiments is problematic. The observation window is typically 100 μs, but lifetimes for SF_6^{-*} can reach into the tens of ms.

We have addressed the previous scarcity of data on pressure dependences in two ways. Experimentally, we measured branching fractions as a function of pressure at multiple high temperatures where appreciable dependences were expected in the pressure range accessible in the FALP. We also confirmed the 1 Torr temperature dependence of branching fractions, as observed by Fehsenfeld at lower temperatures[18]. In addition, statistical modeling of the dependences were carried out with adjustable parameters fixed to a small subset of the data. Modeling results were given for the fraction of SF_5^- ion product as functions of gas temperature, electron temperature, and pressure, including collision-free conditions, in figures 6, 7, 9, and 10 of Part II of the earlier series of papers.[2]

Here we show in figure 6 modeling results for the SF_5^- branching ratio as a function of pressure at various high temperatures, for $T_{el} = T_{gas}$. Our experimental results agree well in the range where thermal detachment (see Sect. V) does not interfere; for simplicity, those data are not shown (see figure 3 of Ref. 2). As expected, temperature and pressure work in opposite ways for SF_5^- production. Increasing pressure increases the fraction of SF_6^- through stabilization of the excited intermediate, and increasing temperature increases the SF_5^- yield since the dissociation rate increases dramatically with internal energy of the SF_6^-. Examining the data, one finds pressure increases in importance with increasing temperature. This is because an increase in the internal energy of SF_6

reduces the lifetime of SF_6^{-}*, which in turn causes the reaction to move further from the high pressure limit into the falloff region. Experiments in the mTorr region would be informative but unfortunately are not available. At temperatures of 560 K and above, the branching data were found to depend on the residence time in the FALP, due to thermal reactivation of stable SF_6^- and subsequent detachment, i.e., thermal detachment. No other ion products other than SF_5^- and SF_6^- were observed at temperatures up to 670 K.

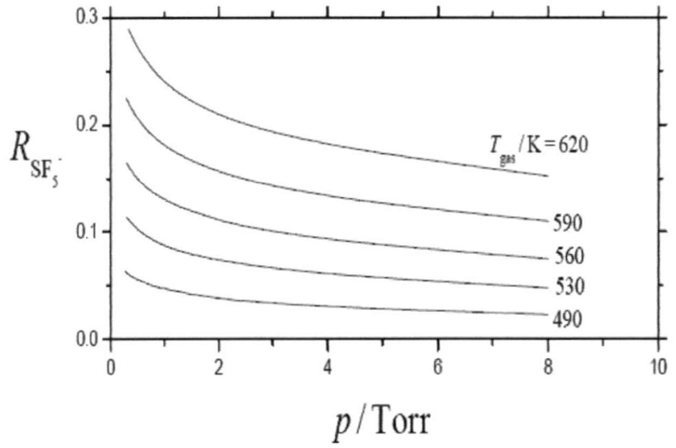

Figure 6. Model results for SF_5^- branching fractions for dissociative attachment under collision-free conditions.

One particularly interesting aspect of the data is that at 590 K the branching fraction reaches the high pressure limit and that significant quantities of SF_5^- are still produced, i.e., typical Stern-Volmer behavior is not followed. Modeling by a simple stepladder analysis, which involves only collisions that reduce the energy of SF_6^{-}*, cannot account for this behavior. Instead, we were forced to use the more accurate and time consuming master equation method, which allows for both energy increasing (up) and energy decreasing (down) collisions. Without the inclusion of up collisions, all SF_6^{-}* would eventually be quenched and no SF_5^- would be observed at high pressure. The up collisions included in the master equation lead to SF_6^{-}* ions with enough energy to dissociate rapidly, i.e., before stabilization can occur in a second collision.

One outcome of this analysis is to confirm an observation made by Španěl et al.[20] and verified by Smith et al.[24] that the peak in the SF_5^- signal observed in beam experiments around an electron energy of 0.3 – 0.6 eV is due to enhanced dissociation of SF_6^{-}* as a result of vibrational excitation by the incoming electron. Our model of this effect was shown in figure 11 of the earlier Part II paper,[2] for collision-free conditions, with a comparison to data by Braun et al.[25] Figure 7 shows measured and modeled cross sections for SF_5^- production as a function of electron energy. The two-peak nature of the data is well reproduced by the ground state only model. The decrease at low energy essentially follows the SF_6^- yield and results exclusively from vibrationally excited SF_6 thermally populated. The increase observed at 0.1 eV is the result of the steep dependence of k_{dis} on energy shown in figure 5. The data peak when essentially all products are SF_5^- and decrease according to the total cross section decrease.

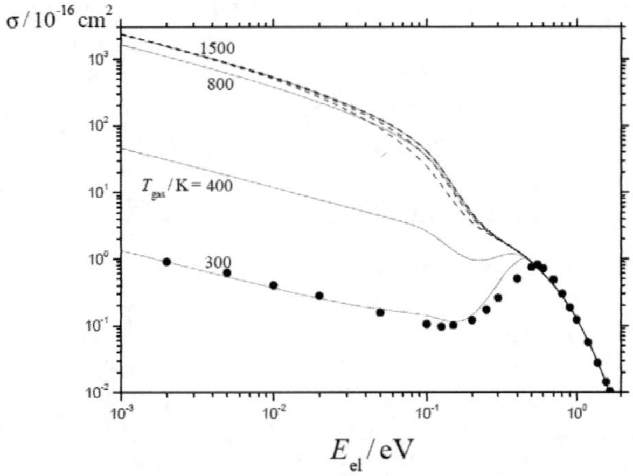

Figure 7. Total (---) and dissociative (——) cross sections as a function of T_{gas} and E_{el} (lines = modeling from this work for T_{gas} as indicated, see text; points = experimental data from Ref. 25 for T_{gas} = 300 K.

V. Autodetachment and Thermal Detachment

Hidden in the above discussion has been the importance of detachment processes at high temperatures. Branching fraction data obtained at high temperature with the FALP showed the importance of thermal detachment. In this process, energy is transferred to stable SF_6^- in collisions with the buffer gas. The SF_6^{-}* that is formed has enough energy to detach the electron,

$$SF_6^- + M \rightarrow SF_6^{-*} + M \qquad \textit{thermal activation.} \qquad (9)$$

Since dissociation occurs at substantially higher energies than detachment, and the SF_6^{-}* is formed near threshold, thermal detachment dominates over thermal dissociation, and the latter is not discussed further. Champion et al.[26] have shown that at higher energies dissociation becomes significant, as indicated by the relative k(E,J) curves for detachment and dissociation. Thermal detachment has been mainly ignored previously although it was looked for at temperatures below 600 K and not observed (see Christophorou and Olthoff[4] and Španěl et al.[20]) In order to determine thermal detachment rates, the SF_6^-/SF_5^- ratio at fixed temperature, pressure, electron concentration, and SF_6 concentration was measured as a function of residence time in the FALP apparatus. Since this ratio is extremely sensitive to detachment, rate constants on the order of tens of s^{-1} could be accurately determined. Above 560 K, a detachment rate constant was measurable. SF_6^- detachment rate constants of 22 (590 K), 40 (620 K), 114 (650 K), and 166 s^{-1} (670 K) were measured. Combining the detachment and attachment rate constants measured under the exact conditions allows for determination of the electron affinity EA(SF_6), as discussed in the following section.

VI. Thermodynamics

There are two important thermodynamic quantities needed in the calculations and the new data allow for accurate determinations of both the electron affinity of SF_6 and the bond dissociation energy D_0^o(F—SF_5^-). Recent reviews of these quantities are found in Christophorou and Olthoff,[5] Miller et. al.[27] and Lobring et al.[28] EA(SF_6) has a long history of being difficult to measure in spite of numerous efforts. This is in part due to the large bond length change in going from SF_6 to SF_6^-, though both have octahedral symmetry. Therefore, the Franck Condon overlap is poor, and finding the origin band corresponding to the EA, in photoelectron spectra, is difficult if not impossible.[29] Similarly the competition between SF_5^- and e^- production makes the determination of the bond energy difficult experimentally.

In the present study, we have measured rate constants for both attachment and detachment under exactly the same conditions. Combining these yields the the equilibrium constant,

$$e^- + SF_6 \Leftrightarrow SF_6^-;\ K = k_{at}/k_{det} \qquad \textit{equilibrium.} \qquad (10)$$

From the equilibrium constants and calculated frequencies for SF_6 and SF_6^-, by a third-law analysis the electron affinity is found to be 1.20 (± 0.05) eV.[2] There have been numerous previous determinations of the electron affinity with no general consistency in the values. A recent review[5] lists two similar values as the preferred choice[30,31] but one of those is from interpretations[31] known to have problems (see Ref. 32). Therefore, the charge transfer equilibrium value[30] of 1.05 (± 0.10) eV has been considered the established value. The present value for EA(SF_6) is 0.15 eV larger than this, just at the limits of the combined uncertainties. Our data do not seem compatible with the lower value. A value EA(SF_6) = 1.05 eV would lead to detachment rate constants 15 times larger than the measured ones at 650 K, which would have resulted in quite obvious upcurving of FALP electron density decay plots, with detachment overwhelming diffusive loss of electrons. While the FALP determination of EA(SF_6) is based mainly on the ratio of k_{at} to k_{det}, it also involves adjustments due to the entropy and heat capacity changes in going from anion to neutral SF_6. The detachment rate constant k_{det} may be written as[33]

$$k_{det} = k_{at}\, N_A\, (273.15/T)\, \exp[-(EA/kT)-(\Delta S^o/k)-(H_T-H_0)/kT]. \qquad (12)$$

In equation (2), k is Boltzmann's constant, N_A is the Avogadro constant, EA is the electron affinity (at 0 K, by definition), ΔS^o is the entropy change due to electron attachment at temperature T, and H_T-H_0 is a thermal energy correction from the measurement temperature to 0 K. Equation (11) is written to emphasize that the thermal electron detachment rate constant is not solely a function of EA and T, but also depends on the entropy change at temperature T and on a heat capacity term. The adjustments are calculated based on neutral and anion symmetries, rotational constants and vibrational frequencies and, in other cases we have studied, affect EA from <1% to nearly 9%.[34] In the case of SF_6 the calculated adjustments amount to 0.1 eV or 9%, i.e., they are significant. This point is emphasized because the SF_6 anion frequencies are not as well established as are those for the neutral, and require a high-level calculation to achieve the desired accuracy; see Ref. 3.

The bond dissociation energy for SF_6^- is determined from a detailed fitting of the SF_5^-/SF_6^- ratio under well defined conditions, both thermal and those taken as function of electron energy. Only a narrow range of D_{298}^o(F—SF_5^-) gives good fits to a variety of data. A value of 1.61 (± 0.05) eV reproduces the data well. Two recent values for this quantity are 1.32 eV by Španěl et al.[20] (adjusting by the new EA) and <1.85 eV by Lobring et al.[28] The former is derived from simple RRK modelling of the SF_5^- branching rates. The detailed modelling present here should supplant that. The latter is derived from threshold measurements of SF_5^- production in the collision induced dissociation of SF_6^-. The less-than sign was attributed to a barrier in the reaction, but may be due to a neglect of the collisional electron detachment channel. We find no barrier need be postulated, in agreement with Španěl et al.[20] Table 1 in Lobring et. al.[28] shows a variety of computational values for D_0^o(F—SF_5^-). Values range from 1.51-1.60 eV from compound Gaussian methods, 1.52 eV from a coupled cluster calculation, and 1.01-2.11 eV from a study using density functional theory. Not included is a value of 1.65 eV from the compound G3 method.[27] All of these computational ranges are consistent with our determination of D_0^o(F—SF_5^-) = 1.61 (± 0.05) eV.[2]

VII. Lifetimes

Lifetime measurements for SF_6^{-}* have been made over a number of years with poor agreement. For instance, the two most recent measurements are 19.1 μs and 1 to >10 ms (Refs. 35 and 36, respectively). Typically, beam experiments find short lifetimes and trap experiments long lifetimes that depend on storage time (see Liu et al.[36]). Our results address this discrepancy.

Lifetimes and $k_{det}(E,J)$ curves (see figure 5) are the inverse of each other (possibly including integration). Our results show that the lifetime will strongly depend not only on the total energy, E, but also the total angular momentum J. For instance at, $k_{det}(1.4 \text{ eV}, J = 0) \approx 10^4 \text{ s}^{-1}$ and $k_{det}(1.4 \text{ eV}, J = 100) = 60 \text{ s}^{-1}$. The latter is about the same at $k_{det}(1.25 \text{ eV}, J = 0)$. Obviously the exact formation mechanism matters greatly. For thermal processes, the $k_{det}(E,J)$ used in the present calculations are more consistent with the trap measurements. In this range, radiative stabilization can compete with detachment. Higher-energy SF_6^{-}* will dissociate faster leaving species with longer lifetimes. Both processes will tend to lead to lifetime measurements that increase with time as have been observed.

We postulate several ideas for the large disparity in lifetimes. The SF_6^{-*} formation process needs to be well defined so that the energy and angular momentum of the complex is well known. If any collisions take place after formation, the lifetime will change dramatically because of the steep energy dependence of the specific rate constant $k_{det}(E,J)$, see figure 5. They also have significant structure around threshold, as shown in figure 6 of our earlier paper,[1] so that the energy distribution of the electron source becomes an important consideration. Assuming lifetimes ~10 ms and a collisional value for the energy transfer rate constant (~10^{-9} cm^3 s^{-1}), one finds number densities as low as ~10^{11} cm will alter the lifetime. An order of magnitude lower pressure still effects 10% of the SF_6^{-}* molecules. In supersonic beams with ill-defined SF_6 temperatures, lifetimes are effected by both E and J dependences. This results both from the $k_{det}(E,J)$ for detachment and for IVR since IVR increases with increasing temperature. The lack of IVR would leave the complex as $e^{-}SF_6$ rather than SF_6^{-}* which would decrease the lifetime. This may be an issue in the recent determination by Le Garrec et al.[35] A potential problem not discussed previously in lifetime measurements is that large electric fields can strip electrons off highly excited molecules. In zero electron kinetic energy (ZEKE) experiments this is a well known issue and electron detachment occurs even at energies slightly below threshold. Large extraction fields may then lead to shorter apparent lifetimes.

VIII. Remaining Issues

It is worth mentioning issues that are unresolved. First is the question of whether F^{-} should be observable in attachment to SF_6 with zero-energy electrons. G3 theory gives an endothermicity of 1.18 eV for this channel at 298 K.[27] The NIST WebBook gives data which place this channel at 0.55 eV endothermic, the difference being in a low value of D(SF_5-F) used by NIST, a value which was reexamined by Tsang and Herron.[37] Their bond strength is consistent with the result of G3 calculations. Braun et al.[25] see a very weak F^{-} ion signal, 3-4 orders of magnitude lower than the SF_6^{-} signal. They left open the possibility that the F^{-} was due to impurities, and planned further work. Because of the report of Braun et al., we used the FALP to search for an F^{-} signal in attachment to SF_6 at 585 K, and place a limit of <0.02% for the branching fraction for F^{-} based on the width of our baseline. This limit includes a mass discrimination correction which was uncovered in recent FALP work.[38]

Earlier we referred to results on photoelectron spectroscopy of SF_6^{-}.[29] That work showed a resolved, long progression of vibrational excitation in the ν_1 symmetric stretch mode of SF_6 as a result of detachment of the extra electron. It also showed a companion peak which was unexplained until recently, as simultaneous excitation of ν_1 and ν_4 modes.[39] Reference 29 also contained infrared vibrational predissociation spectra for SF_6^{-}, in which a resolved ν_1 transition was observed. Those spectra also contained structure, which is unexplained as of this writing. Thus, not everything may be known about the vibrational structure of SF_6^{-}.

IX. Conclusions

Experimental studies were conducted on electron attachment to SF_6 and on thermal electron detachment from SF_6^- to provide input for statistical modeling of the extensive body of data on the electron-SF_6 reaction system at low electron energies. In particular, the pressure dependence of attachment and detachment were examined, because little attention has been paid to collisional stabilization in the past. The details of the modeling were presented in earlier papers.[1-3] Empirical rate constants for the various processes (figure 1) involved in the electron attachment and stabilization problem were used to predict temperature, pressure, and electron temperature dependence of the attachment and dissociation (into F + SF_5^-) processes. Most absolute electron attachment experiments carried out to date have been in the high-pressure limit, where collisional equilibration occurs between SF_6^- and SF_5^-. The present modeling shows that the high-pressure limit moves to higher pressures as the gas temperature is increased. Aside from the new attachment and detachment rate constants measured, the main conclusions which may be drawn from this work are (1) only the ground electronic state of SF_6^- need be invoked to explain available data; (2) the electron affinity of SF_6 is higher than previously thought – we find EA(SF_6) = 1.20 (± 0.05) eV; (3) the endothermicity of the dissociative electron attachment reaction that yields SF_5^- is 0.41 (± 0.05) eV at 0 K; (4) combining these two numbers gives the bond energy D_0^o(F—SF_5^-) = 1.61 (± 0.05) eV.

The modeling of the specific rate constants of the competing processes is, of course, dependent upon the quality of the experimental data. While there are a number of experimental data sets that are in agreement on, for example, the attachment rate constant in the high-pressure limit, other vital data are the product of single reports. It would be most helpful if future experiments could be carried out to (a) pin down EA(SF_6) more accurately, (b) confirm the low temperature results of Ref. 12, (c) confirm the very low-pressure ICR work of Ref. 15 and provide pressure dependences for electron attachment in the μTorr to mTorr region, and (d) provide additional thermal electron detachment data for SF_6^-. We are hoping to be able to extend some of our own measurements to higher temperatures at some point in the future.

X. Acknowlegments

This research was funded by the United States Air Force Office of Scientific Research under Project 2303EP4. Encouragement of this work by M. Berman, helpful discussions with H. Hotop and E.E. Nikitin, technical assistance by A. I. Maergoiz, as well as support by the Deutsche Forschungsgemeinschaft (SFB 357 "Molekulare Mechanismen unimolekularer Prozesse") are gratefully acknowledged.

References

[1] Troe J, Miller T M and Viggiano A A 2007 *J. Chem. Phys.* **127** 244303.
[2] Troe J, Miller T M and Viggiano A A 2007 *J. Chem. Phys.* **127** 244304.
[3] Viggiano A A, Miller T M, Friedman J F and Troe J 2007 *J. Chem. Phys.* **127** 244305.
[4] Christophorou L G and Olthoff J K 2004 *Fundamental Electron Interactions with Plasma Processing Gases* (New York: Kluwer Academic/Plenum).
[5] Christophorou L G and Olthoff J K 2000 *J. Phys. Chem. Ref. Data* **29** 267-327.
[6] Klots C E 1976 *Chem. Phys. Lett.* **38** 61-64.
[7] Hotop H, Ruf M-W, Allan M and Fabrikant I I 2003 *Adv. At. Mol. Opt. Phys.* **49** 85-216.
[8] Dashevskaya E I, Litvin I, Nikitin E E and Troe J (in press) *Phys. Chem. Chem. Phys.*
[9] Braun M, Barsotti S, Marienfeld S, Leber E, Weber J M, Ruf M-W and Hotop H 2005 *Eur. Phys. J. D* **35** 177-191.
[10] Crompton R W and Haddad G N 1983 *Aust. J. Phys.* **36** 15-25.
[11] Petrovic Z L and Crompton R W 1985 *J. Phys. B* **17** 2777-2791.
[12] LeGarrec J L, Sidko O, Queffelec J L, Hamon S, Mitchell B A and Rowe B R 1997 *J. Chem. Phys.* **107**, 54-63.

[13] Goulay F, Rebrion-Rowe C, Carles S, Le Garrec J L and Rowe B R 2004 *J. Chem. Phys.* **121,** 1303-1308.
[14] Christophorou L G and Datskos P G 1995 *Int. J. Mass Spectrom. Ion Processes* **149/150** 59-77 (1995).
[15] Foster M S and Beauchamp J L 1975 *Chem Phys. Lett.* **31** 482-486.
[16] Fernandez A I, Viggiano A A, Miller T M, Williams S, Dotan I, Seeley J V and Troe J 2004 *J. Phys. Chem. A* **108** 9652-9659.
[17] Dunbar R C, (personal communications).
[18] Fehsenfeld F C 1970 *J. Chem. Phys.* **53** 2000-2004.
[19] Rains L J, Moore H W and McIver R T 1978 *J. Chem. Phys.* **68** 3309-3311.
[20] Španěl P, Matejcik S and Smith D 1995 *J. Phys. B* **28** 2941-2957.
[21] Chen C L and Chantry P J 1979 *J. Chem. Phys.* **71** 3897-3907.
[22] Ferguson E E 1972 in *Ion-Molecule Reactions*, vol. 2, Ed. J. L. Franklin (London: Butterworth), chap. 8, p. 370.
[23] Hickam W M and Berg D 1958 *J. Chem. Phys.* **29** 517-523.
[24] Smith D, Španěl P, Matejcik S, Stamatovic A, Märk T D, Jaffke T and Illenberger E 1995 *Chem. Phys. Lett.* **240** 481-488.
[25] Braun M, Marienfeld S, Ruf M-W and Hotop H (to be published).
[26] Champion R L, Dyakov I V, Peko B L and Wang Y 2001 *J. Chem. Phys.* **115** 1765-1768.
[27] Miller T M, Arnold S T and Viggiano A A 2003 *Int. J. Mass Spectrom.* **227** 413-420.
[28] Lobring K C, Check C E, Gilbert T M and Sunderlin L S 2003 *Int. J. Mass Spectrom.* **227** 361-372.
[29] Bopp J C, Roscioli J R, Johnson M A, Miller T M, Viggiano A A, Villano S M, Wren S W and Lineberger W C 2007 *J. Chem. Phys.* **111** 1214-1221.
[30] Grimsrud E P, Chowdhury S and Kebarle P 1985 *J. Chem. Phys.* **83** 1059-1068.
[31] Chen E C M and Chen E S 2004 *J. Chromatography A* **1037** 83-106.
[32] Ervin K M, Anusiewicz W, Skurski P, Simons J and Lineberger W C 2003 *J. Phys. Chem. A* **107** 8521-8529.
[33] Miller T M, Friedman J F and Viggiano A A 2004 *J. Chem. Phys.* **120** 7024-7028.
[34] Miller T M 2005 *Adv. At. Mol. Opt. Phys.* **51** 299-342.
[35] Le Garrec J-L, Steinhurst D A and Smith M A 2001 *J. Chem. Phys.* **114** 8831-8835.
[36] Liu Y, Suess L and Dunning F B 2005 *J. Chem. Phys.* **122** 214313.
[37] Tsang W and Herron J T 1992 *J. Chem. Phys.* **96** 4272-4282.
[38] Van Doren J M, Miller T M, Viggiano A A, Španěl P, Smith D, Bopp J C and Troe J (in press) *J. Chem. Phys.*
[39] Borrelli R 2007 *Chem. Phys. Lett.* **445** 84-88.

Journal of Physics: Conference Series **115** (2008) 012020 doi:10.1088/1742-6596/115/1/012020

Electron Collisions with Biomolecules

Vincent McKoy and Carl Winstead

California Institute of Technology, Pasadena, California 91125, USA

mckoy@caltech.edu

Abstract. We report on results of recent studies of collisions of low-energy electrons with nucleobases and other DNA constituents. A particular focus of these studies has been the identification and characterization of resonances that play a role in electron attachment leading to strand breaks in DNA. Comparison of the calculated resonance positions with results of electron transmission measurements is quite encouraging. However, the higher-lying π^* resonances of the nucleobases appear to be of mixed elastic and core-excited character. Such resonant channel coupling raises the interesting possibility that the higher π^*resonances in the nucleobases may promote dissociation of DNA by providing doorway states to triplet excited states.

1. Introduction

The demonstration by Sanche and coworkers [1,2,3,4] that low-energy electrons can induce single- and double-strand breaks in DNA has stimulated significant interest in the interactions between slow electrons and the molecular constituents of DNA and RNA. Because the observed yield of strand breaks exhibits peaks as a function of incident electron energy, processes involving resonances in the electronically elastic channel are implicated, particularly resonance-enhanced dissociative attachment. Not surprisingly, there have been numerous experimental studies of dissociative attachment to the nucleobases [5,6,7,8,9], which have not only demonstrated that resonant dissociative attachment indeed occurs in these bases but have also provided the mechanistic insight that gas-phase dissociative attachment may be driven largely by vibrational Feshbach resonances built on dipole-bound temporary anions. There have also been measurements of the derivative electron-transmission spectra of the nucleobases and their halogenated derivatives that reveal the location of the low-energy scattering resonances [7,10]. Comparison between the attachment and transmission measurements on the nucleobases also indicates that the strongest resonance features in the total cross section (presumably the π^* elastic shape resonances) are not necessarily responsible for the strongest features in the dissociative-attachment cross section [3,7,11]. These experimental studies have been complemented by several calculations of electron-collision cross sections for the nucleobases [12,13,14,15,16,17].

In this contribution we review the results of our calculations of low-energy electron collisions with the nucleobases and other DNA constituents. Our studies have concentrated on the identification and characterization of low-energy shape resonances that play a role in the initial capture electrons and thus in promoting subsequent disruption of the DNA structure, as well as on examining how the resonance positions change when smaller subunits are incorporated into larger moieties. On balance, our calculations support the assignments by Burrow and coworkers [7,10] of the electron-transmission features to π^* shape resonances and thus support the interpretation of the main dissociative-attachment

Journal of Physics: Conference Series **115** (2008) 012020 doi:10.1088/1742-6596/115/1/012020

peak as arising from an alternative mechanism, namely a vibrational Feshbach resonance. A consistent discrepancy between the calculated and measured positions for the third π^* resonance in each nucleobase probably reflects the influence of channel-coupling effects not included in our calculations. That is, these higher-lying resonances are not pure elastic shape resonances but also have core-excited character built on the low-lying triplet excited states into which they may decay. This raises the interesting possibility that the higher π^* resonances in the nucleobases may promote dissociation of DNA by providing doorways to triplet excited states. The significance of such resonant mixing is demonstrated for pyrazine, a high-symmetry analogue to the pyrimidine bases.

2. Theoretical and computational aspects

Our work employs the Schwinger multichannel (SMC) method [18]. The SMC method is a full many-electron formulation of electron-molecule collisions that is capable of treating the complications arising in low-energy collisions, including exchange, polarization, and electronic excitation. As with other high-level methods that aim to treat the full complexity of the electron-molecule collision process [19,20], the computational demands of the SMC method scale rapidly with the size of the molecule of interest and can become extreme for large polyatomic molecules. Key challenges in our work have therefore been the development of computer codes that run efficiently on parallel computers and workstation clusters [21,22] and improvements in the design of calculations that accelerate convergence and thus mitigate the scaling problem. By formulating the SMC method for massively parallel computers [21], we have been able to achieve sustained performance in the hundreds of gigaflops, making studies of electron collisions with targets such as the nucleobases and deoxynucleosides feasible.

3. Results

Over the past few years, we have employed the parallel implementation of the SMC method to carry out all-electron calculations of the collisions of low-energy electrons with the 5 nucleobases as well as with deoxynucleosides and the deoxynucleotide deoxyadenosine 5′-monophosphate [14,15,16]. In these calculations, exchange effects were treated explicitly and polarization effects were included through virtual excitations of the molecule (closed channels) when computing the π^* resonance energies. Though the nucleobases are not quite planar, each molecule was treated in C_s symmetry. Imposing a planar geometry on these molecules facilitates the computations and their analysis by allowing us to separate the orbitals and electronic states into either the $^2A''$ representation or $^2A'$ representation; in particular, the π^* resonances fall into the $^2A''$ representation where they can be more readily distinguished from the large $^2A'$ background. However, distorting the molecular geometry may shift resonance positions, as we recently found, for example, in tetrahydrofuran [23], but these shifts are expected to be small for the nucleobases. In fact, calculations of the static-exchange cross sections at the C_s and undistorted (C_1) geometries for the purine base guanine determined that the π^* resonances are shifted up by about 0.2 eV in the planar configuration. Furthermore, the nucleobases are all polar, and long-range scattering of electrons by the dipole potential leads to strong forward scattering that can be difficult for methods that rely on finite basis sets, such as the SMC method, to capture. Though there are procedures for correcting the calculated cross sections to account for this long-range scattering, we have neglected such corrections because they are not expected to significantly affect the π^* resonance energies of interest here.

Fig. 1 shows our computed $^2A''$ cross section for the purine base adenine [15], displaying the low-energy resonances associated with temporary trapping of the electron in vacant π^* orbitals. We show results only up to 5 eV because of pseudoresonances at higher energy which make it difficult to determine the actual location of the fourth resonance. Our computed resonance energies of 1.1, 1.8, and 4.1 eV compare reasonably well with the resonances observed at 0.54, 1.36, and 2.17 eV in transmission measurements [10]. The agreement between the computed resonance positions and the transmission measurements for the other nucleobases, uracil [14[, thymine and cytosine [16], and guanine [15], is quite similar to that seen here for adenine, and we may conclude that our calculations support the assignment of the observed electron-transmission resonances [7,10] as π^* resonances. In contrast, other scattering calculations [12,13,17] place the π^* resonances considerably higher in energy, requiring either empirical shifts [13] or the association of the first calculated π^* resonance with the second resonance observed in the transmission spectra [12,24]. The calculations of Tonzani and Greene employ local approximations to both the exchange and polarizations potentials, and, while both approximations may introduce errors, errors due to the approximation to exchange are probably responsible for much of the error in the resonance energies, because comparable errors are seen when comparing a static-exchange calculation for CO_2 using the same procedure [25] with all-electron static-exchange calculations.

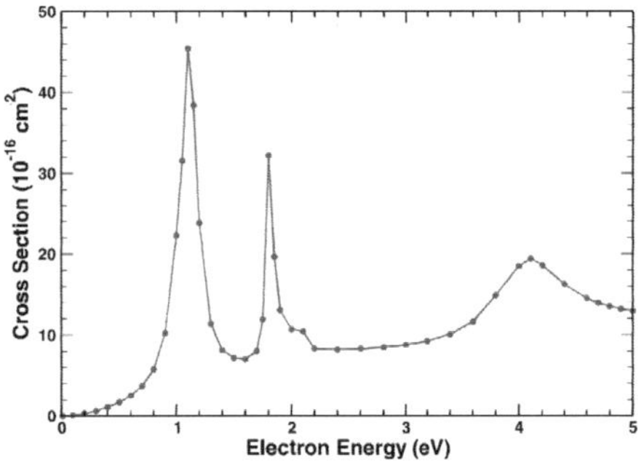

Figure 1. *A″* component of the integral elastic cross section for electron scattering by adenine as computed in the static–exchange plus polarization approximation. The three prominent peaks are due to π^* resonances.

The error of the order of 2 eV between our computed position for the third π^* resonance in adenine and the electron-transmission measurement is much less satisfactory than for the two lowest-energy π^* resonances where the errors are about half an eV. We have observed this pattern consistently in our calculations on the nucleobases [14,15,16] where the calculated positions of the two lowest and narrowest resonances agree quite well with the measurements but the computed locations of the third and broadest resonance are consistently too high. While these discrepancies may partly

arise from limitations of our treatment of polarization, it was surprising that the error is largest for the highest-energy and broadest resonance, because polarization effects are expected to be less important as the collision time decreases. In [15] we surmised that this larger discrepancy was due to some underlying difference in the physical character of the higher-lying π^* resonances from that of the lower resonances that was not being captured by our choice of configurations describing polarization effects.

In our calculations of electron collisions with nucleobases polarization effects were represented by including terms built on singlet virtual excitations of the molecule in the scattering wavefunction. Such excitations describe the relaxation of the molecular charge density and account for the long-rang polarization interactions as well as shorter-range electron correlation effects. By excluding configurations built on triplet excited states of the molecule, we precluded the possibility of coupling between the elastic-channel shape resonances and core-excited resonances built on low-lying triplet states of the nucleobases. Such resonant channel coupling might be expected to be more important for the higher-lying resonances than for those lying well below the lowest triplet excitation threshold and could account for the trend we observed in our calculated π^* resonance positions in the nucleobases. Moreover, because such mixed resonances can decay not only into the elastic channel but also into triplet states, they may play a role in promoting DNA damage by slow electrons.

To explore such channel mixing [26,27] we chose pyrazine, a close analogue of the pyrimidine bases, because of its high symmetry (D_{2h}) and the experimental data available on the positions of its three low-energy resonances [28]. Calculations [26,27] reveal that the two lowest resonances are nearly pure single-channel resonances but the third is, indeed, as long suspected [28], heavily mixed with core-excited resonances built on triplet $\pi \rightarrow \pi^*$ excited states. In fact, Nenner and Schulz [28] posited that the higher-lying shape resonances they had observed in benzene and azabenzenes were probably mixed with core-excited shape or Feshbach resonances associated with low-lying excited states of the molecule. Though such mixing was confirmed by subsequent observation in benzene of the decay of the higher π^* resonance into both the elastic channel and a triplet excited state [29], it had not been previously accounted for in calculations on elastic electron-molecule collisions nor had its potential to produce a large and selective effect on resonance energies been generally appreciated.

To make a closer connection between our results and DNA itself, we have also studied collisions of slow electrons with the purine and pyrimidine nucleosides and the deoxynucleotide 2′-deoxyadenosine 5′-monophosphate [15,16]. For these systems, higher-level calculations including polarization effects are not yet feasible with the current versions of our computer codes, and they were therefore studied at the static-exchange level. Though the static-exchange cannot be expected to yield correct resonance positions, it is fairly reliable above 10 eV, where polarization effects are small, and provides upper bounds to resonance energies because the omitted polarization interaction is attractive. Moreover, by comparing resonance positions obtained with and without polarization on the nucleobases, we can make informed estimates of the actual resonance positions in the deoxynucleosides and deoxynucleotides.

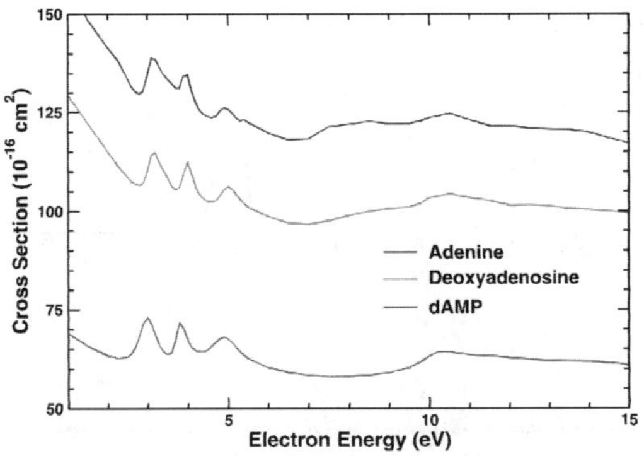

Figure 2. Integral elastic cross sections for electron scattering by adenine (bottom, red), 2-deoxyadenosine (middle, green), and 2-deoxyadenosine 5-monophosphate (top, blue), computed in the static--exchange approximation.

Fig. 2 shows our calculated integral elastic cross sections for collisions of low-energy electrons with adenine, its nucleoside, deoxyadenosine, and its nucleotide 2′-deoxyadenosine 5′-monophosphate (dAMP). For comparison, the adenine cross sections in this figure were also obtained at the static-exchange level which neglects polarization effects. The significant upward shift in the π^* resonance positions in the nucleobase in comparison to the results of Fig. 1 reflects the importance of polarization effects. Details of the calculations, including the assumed nuclear geometries, can be found in [15]. Because of the lack of symmetry in the larger molecules, we can no longer separate the π^* resonances cleanly from the background, as we could in adenine, but their positions are still evident. There is clearly a close correlation between the low-energy resonances of the nucleoside and those of the nucleobase, the main difference being a slight upward shift of the resonances in deoxyadenosine relative to adenine. The positions of the π^* resonances, however, are virtually unchanged going from the nucleoside to the nucleotide (dAMP).

4. Summary

We have reviewed results of first-principles calculations of the collisions of slow electrons with the nucleobases of DNA and RNA as well as with the nucleosides and nucleotides of the DNA bases [14,15,16]. Our studies have emphasized the characterization of the π^* resonances in the nucleobases and our calculated positions for these resonances support the assignments proposed by Burrow and coworkers [7,10] on the basis of electron transmission measurements. In the course of our studies of the nucleobases, we uncovered an interesting difference between the lowest-energy π^* resonances and those at higher energy. Exploring this difference in the high-symmetry model compound pyrazine, a close analogue of the pyrimidine bases, we found that the third π^* resonance actually has mixed character: it is partly an elastic channel shape resonance and partly a core-excited resonance built on low-lying triplet excited states, and inclusion of this resonant channel coupling is essential to obtaining the correct resonance energies. This result raises the interesting possibility that the higher π^* resonances (and perhaps other high-energy shape-resonances) in the nucleobases may promote dissociation of DNA by providing doorways to triplet excited states.

Acknowledgments

This research was supported by the Chemical Sciences, Geosciences and Biosciences Division, Office of Basic Energy Sciences, Office of Science, U.S. Department of Energy. We are grateful for the use of the resources of the Jet Propulsion Laboratory's Supercomputing and Visualization Facility.

References

[1] Boudaïffa B, Cloutier P, Hunting D, Huels M A and Sanche L 2000 *Science* **287** 1658
[2] Huels M A, Boudaïffa B, Cloutier P, Hunting D and Sanche L 2003 *J. Am. Chem. Soc.* **125** 4467
[3] Martin F, Burrow P D, Cai Z, Cloutier P, Hunting D and Sanche L, 2004 *Phys. Rev. Lett.* **93** 068101
[4] Panajotovic R, Martin F, Cloutier P and Sanche L, 2006 *Radiat. Res.* **165** 452
[5] Huels M A, Hahndorf I, Illenberger E and Sanche L 1998 *J. Chem. Phys.* **108** 1309
[6] Abdoul-Carime H, Huels M A, Illenberger E and Sanche L 2001 *J. Am.Chem. Soc.* **123** 5354
[7] Scheer A M, Aflatooni K, Gallup G A and Burrow P D 2004 *Phys. Rev. Lett.* **92** 068102
[8] Denifl S, Ptasi ska S, Probst M, Hrušak M, Scheier P and Märk T D 2004 *J. Phys. Chem.* A **108** 6562
[9] Aflatooni K, Scheer A M and Burrow P D 2005 *Chem. Phys. Lett.* **408** 426
[10] Aflatooni K, Gallup G A and Burrow P D 1998 *J. Phys. Chem.* A **102** 6205
[11] Martin F, Burrow P D, Cloutier P, Hunting D and Sanche L 2004 *Phys. Rev. Lett.* **93** 068101
[12] Gianturco F A and Lucchese R R 2004 *J. Chem. Phys.* **120** 7446
[13] Tonzani S and Greene C H 2006 *J. Chem. Phys.* **124** 054312
[14] Winstead C and McKoy V 2006 *J. Chem. Phys.* **125** 174304
[15] Winstead C and McKoy V 2006 *J. Chem. Phys.* **125** 244302
[16] Winstead C, McKoy V and Sanchez S d'A 2007 J. *Chem. Phys.* **127** 085105
[17] Yalunin S and Leble S B 2007 *Eur. Phys. J. Special Topics* **144** 115
[18] Takatsuka K and McKoy V 1984 *Phys. Rev* A **30** 1734
[19] Burke P G and Berrington K A (eds) 1993 *Atomic and Molecular Processes: An R-Matrix Approach* (Bristol: IOP Publishing)
[20] Rescigno T N, McCurdy C W, Orel A E and Lengsfield B H 1995 *Computational Methods for Electron-Molecule Collisions* ed F A Gianturco and W M Huo (New York: Plenum) p 1
[21] Winstead C and McKoy V 1996 Adv. At. Mol. Opt. Phys. **36** 183
[22] Winstead C and McKoy V 2000 *Comput. Phys. Commun.* **128** 386
[23] Winstead C and McKoy V 2006 *J. Chem. Phys.* **125** 07430

[24] Burrow P D 2005 *J. Chem. Phys.* **122** 087105
[25] Tonzani S and Greene S H 2005 *J. Chem. Phys.* **122** 014111
[26] Winstead C and McKoy V 2007 *Phys. Rev. Lett.* **98** 113201
[27] Winstead C and McKoy V 2007 *Phys. Rev.* A **76** 012712
[28] Nenner I and Schulz G J 1975 *J. Chem. Phys.* **62** 1747
[29] Allan M 1982 *Helv. Chim. Acta* **65** 2008

AUTHOR INDEX